THINKING THROUGH
THE IMAGINATION

AMERICAN PHILOSOPHY

Douglas R. Anderson and Jude Jones, series editors

THINKING THROUGH THE IMAGINATION

Aesthetics in Human Cognition

JOHN KAAG

FORDHAM UNIVERSITY PRESS   NEW YORK   2014

Fordham University Press has no responsibility for the persistence or accuracy of URLs for external or third-party Internet websites referred to in this publication and does not guarantee that any content on such websites is, or will remain, accurate or appropriate.

Fordham University Press also publishes its books in a variety of electronic formats. Some content that appears in print may not be available in electronic books.

Library of Congress Cataloging-in-Publication Data

Kaag, John J., 1979–
    Thinking through the imagination : aesthetics in human cognition / John Kaag. — First edition.
        pages cm — (American philosophy)
        Includes bibliographical references and index.
        ISBN 978-0-8232-5493-4 (cloth)
    1. Imagination (Philosophy) 2. Aesthetics. 3. Cognition. 4. Kant, Immanuel, 1724–1804. 5. Schiller, Friedrich, 1759–1805. 6. Peirce, Charles S. (Charles Sanders), 1839–1914. I. Title.
    BH301.I53K33 2014
    111'.85—dc23

                                                        2013006701

16 15 14   5 4 3 2 1

First edition

Contents

Acknowledgments

Like most people, I became acquainted with the imagination at a rather early age. In my case, the meeting took place as a child in my mother's backyard garden. My mother ensured that my contact with the imagination would not be a passing acquaintance. She taught her children to acknowledge and actualize the possibilities that life affords even, and perhaps especially, when they were not readily apparent. This was an especially useful lesson for a not-so-young child who hoped to go into the discipline of philosophy, a field that looks a bit barren at first glance. To the extent that this book is a function of my education as a philosopher, and to the extent that my mother and brother urged me to acquire this education, I have them to thank for the writing of this book.

I began this project under the guidance of Douglas Anderson, who, as my mentor and friend at Penn State, encouraged me to roam freely over a variety of academic fields and to focus carefully when the roaming became listless. Mark Johnson, Scott Pratt, and John Lysaker helped me remember that this careful focus on analysis and argumentation could be the stuff of imagination and meaning. This is to say that they were ideal graduate professors. Much that is correct or beautiful about this book I owe to their guidance and willingness to converse; the mistakes and boring bits are, of course, my own. I would like to thank a number of other peers and colleagues who have helped me give shape to this book: Frank Oppenheim, Robert Innis, John McDermott, Joseph Margolis, Erin McKenna, Mat Foust, Kim Garchar, Jeff Downard, Michael Raposa, Roger Ward, and Rob Main. Dawn Aberg's help in proofing and editing the first draft of the volume was extremely helpful. I thank Rogers Hollingsworth and Gerald Edelman for encouraging me at a crucial point in the development of the manuscript, on a spring afternoon when I was very much inclined to draw this analysis of the imagination up short. Much of this research was supported by the American Academy of Arts and Sciences and the Harvard Humanities Center; I thank Patricia Meyer Spacks and David Sehat for their invaluable encouragement and criticism over the course of the research and writing.

Finally, there is a person who does not fall neatly into the category of peer, or friend, or mentor, or supporter, or critic, or family member. Carol Hay plays all of these roles—seamlessly, creatively, joyfully. No other person has done more in helping me think through the imagination, through its risks and potentialities.

THINKING THROUGH
THE IMAGINATION

ONE

THE CULTIVATION
OF THE IMAGINATION

It is not by dealing out cold justice to the circle of my ideas that I can make them grow, but by cherishing them and tending them as I would flowers in my garden.

—Charles Sanders Peirce (1893)

The Imaginative Imperative

For the two children, the season began as a wild dash—a race against the length of summer days.[1] But by mid-August, the days proved too long and hot for their short attention spans. The unconstrained freedom of vacation exhausted itself or, more accurately and more ironically, exposed itself as a type of aimless discontent. Freedom from chores, school, and responsibility revealed itself as boredom to me and my brother on a humid afternoon. The toys and blocks that had once riveted our attention lay thrown and neglected about the playroom. Haphazardly discarded games no longer occupied our full attention. Surrounded by a chaos of playthings, my brother and I sat bickering in the middle of the room. At least bickering gave us something to *do*.

My mother had been listening to us for some time from the garden. The injunction that came to us through the back window was as simple as it was emphatic:

"Boys! Stop Squabbling! *Be Imaginative!*"

Being imaginative is no simple matter for two tired youngsters. More often than not, we needed a bit of encouragement. My brother and I had contented ourselves with our discontent, objecting to any force that might jostle us out of the odd comfort of bickering. Encouragement came in the form of an order. Get up off the ground, pick up our blocks and games, and come outside to help in the garden.

If being imaginative meant helping in the garden, my brother and I wanted no part in it. How could imagination play freely if it was forced to help with mundane chores? Gardening, however, if done properly, is an engaging activity, and our reluctance was short-lived. It is, after all, difficult to be reluctantly imaginative. Planting a bed is a type of play that rarely grows old. After a short tutorial in gardening etiquette, my mother set us free on a small plot. We were, however, not wholly free, at least not in the negative sense of being free *from* school or free *from* chores. This gardening may have been free play, but it was also serious business that deserved our full attention. In being imaginative, my brother and I came to understand the rules of play, the guidelines that determined the arrangement of shrubs and hosta, as they emerged unexpectedly in the interaction with a variety of plants and in a particular garden. This variety was not embodied in the random scattering of discarded toys or blocks but rather in the gathering together of various plants into the felt harmony of a well-planted garden. Even youngsters can learn that such a novel but harmonious gathering is the meaning of being imaginative. I had not yet read John Dewey's *Art as Experience*, of course. But I did in a certain intuitive way understand that

> the imagination is a way of seeing and feeling things as they compose an integral whole. It is the large and general blending of interests at the point where the mind comes in contact with the world. When old and familiar things are made new in experience there is imagination.[2]

The free play of growth and cultivation is a process that involves a child even against his will. Being imaginative means getting your hands dirty. Really dirty. After a stint in the yard, my mother would joke that it was difficult to see where the dirt ended and the skin began. In truth, such distinctions—between the human and the natural—are difficult to make in the midst of imaginative planting. It is here that we get at least a vague sense of the issues that will emerge over the course of this relatively thin book. In its everyday use, the imagination is understood as a creative power—perhaps *the* creative power—by which human beings get on with the meaningful business of living. It is the imagination that allows us to escape the mediocrity of our daily lives, to transcend the self-imposed boundaries—conceptual, personal, and social—that limit our growth. It is the imagination that generates a work of art, and it is the imagination that grants us the ability to interpret artworks. It is the imagination that keeps culture and science "on the move." In short, it is the imagination that makes us fully human.

The American philosophical tradition, with its emphasis on growth, discovery, and human flourishing, has always been, at least indirectly, interested in this power of transcendence. Indeed, this philosophical tradition arguably more than any other has placed the imagination at the center of its account of being human. But here we encounter a problem, one that troubled American thinkers from Ralph Waldo Emerson to Susanne Langer. *Where does the imagination, a creative power that has been regarded as distinctly human, come from? And where exactly does it take us?* In one sense, these questions lead us into a phenomenological account of the experience of creativity. In another, it challenges us to locate the imagination in the workings of the natural world, in the physical natures that each of us possess. We should not be surprised that this has led to a variety of divergent treatments of the subject. Let us begin on the phenomenological side of things. Just as a youngster gets lost in the planting of a garden, in moments of imaginative play we often seem to lose ourselves in our play, our painting, or our experimentation. It seems that we are called out of our habitual ways of being in the world in order to engage with our surroundings in a new way. We are left unsettled, forever changed. We find ourselves anew, surprised to find that world that has been changed by our hands. For the ancients, these moments of ecstasy were attributed to the work of the muses, and it is in this attribution that they answered, or perhaps avoided, the question of the imagination's origin. Perhaps we will find that the ancient Greeks were right in their avoidance of the question, but it seems that much can be said in regard to the way that the chance occurrences we experience in our natural surroundings give us the chance to exercise our creative powers.

Only recently have I looked back on the planting afternoons in my childhood garden and reflected on the meaning of being imaginative. The reflection is my point of departure for the following project. Dewey was right: the imagination brings individuals together and leads them to compose a unified whole. Our garden was living proof of this. After a successful afternoon of planting, the various shrubs and the various members of my family were, for the time being, arranged in a cooperative relation. As Dewey notes, imagination "weds man and nature" but also "renders men (and women) aware of their union with one another in origin and destiny."[3] This "union" could never have been prescribed beforehand, nor could it be repeated in the future. This union was brand-new—novelty at its finest. Old and familiar plots of ground, tamped down by the careless running of little feet, were carefully retilled and given new life on those days of imagination. The regeneration of the garden corresponded with a kind of personal renewal and growth. The communication and cooperation of two brothers were made anew. The boredom of old and familiar patterns of interaction was overcome in the injunction to be imaginative. There was no time for bickering—we were absorbed in the process of planting. In this process, imagination transformed the dispositions of two boys and the world at large, or at least a small corner of my mother's back lawn. In spite of my efforts to verbalize the meaning of these imaginative days, something about this boyhood experience has been lost—and lost forever. The imagination is cultivated in the continuous flow and interaction of sense, emotion, and meaning. One experiences this flow as a pervasive quality of feeling that defies definitive explanation. Any real attempt to describe this pervasive quality results in an admixture of poetry and paradox that still misses the imaginative quality of gardening. As Proust understood, all of the senses are at play in imagination. The ground *smells* warm. The plants *call* to be planted in a certain fashion. The orange and gold of marigolds *feel right* next to the varied hues of zinnias. While such poetic descriptions hint at the holistic quality of the imagination, they do little to describe the process and situation in which *being* imaginative takes place. Without ignoring the phenomenological character of imaginative encounters, I hope to describe the limits and the movement of imagination. Without abandoning the lived experience of the imagination, I aim to theorize about its particular place in the history of thought and its unique outline as a way of thinking. This theorizing must, if it is to be at all successful, in a certain sense return us to the experience of the imagination.

The investigation of the imagination and its origins must remain accountable to the experience of human creativity. Let us return to our central question: Where does the imagination come from? In answering this, we cannot lose the experiential sense of the imagination. This is no easy task. I maintain that a proper answer will lead us not only into the depths of phenomenology but also, simultaneously, into the most recent findings of the empirical sciences.

This may seem to be a surprising claim. What could the empirical sciences have to say about the creative spirit of the imagination? Is there not a danger in reducing the imagination to a set of mechanical processes, reducing the one human power that is, almost by definition, not to be reduced? Yes, there is. I concede that there is a chance that the empirical sciences may have nothing to contribute to our understanding of the imagination, and there is the very real risk of reductionism. That being said, I think it is worth taking this chance and facing this risk. Since Aristotle and Kant, there has been a vague sense that human creativity—in its cultural and artistic splendor—could be traced to the creative processes of nature. With this project, I hope to show how this vague sense might now be articulated in a coherent theory. It currently remains a mere hope expressed as we look forward into an age of science that *does* respond, despite rumors to the contrary, to the longstanding and well-grounded accusations of material reductionism. What we find when we touch on the findings of *today's* science—not the science of fifty or a hundred years ago—is an increasing interest in the indeterminacy, emergence, and spontaneity of natural processes. This interest allows us to investigate a continuity between the novelty of the human imagination and the workings of nature that appear to reflect a corresponding dynamism. The empirical sciences will not tell us how to be imaginative, but they may shed light on the origins and preconditions of this creative power that has long defined what it is to be human. Led by theoretical biology and cognitive neuroscience, the natural sciences are at the very least on their way to developing models that might give us perspectives on the complexity, novelty, and unity of the creative imagination. That is the hope.

This hope comes hand in hand with the belief that by attending to the experience and the origin of the imagination we can respond to the bickering that emanates from a corner of the philosophic playroom. Like all bickering, it stems as much from boredom as it does from any genuine conflict. Thus, the injunction to "be imaginative" is an appropriate one. Much of the squabbling in contemporary philosophy appears to concern the shape and definition of human thinking and human meaning making. An honest observer could see the disagreement as one that emerges as the toys and games of the history of philosophy lose their charm. More accurately, perhaps, the charm of these games has been forgotten. My project, therefore, is an injunction for philosophy to "be imaginative!"—to attend to the embodied, creative, and spontaneous aspects of human thought that the Western philosophical canon does occasionally highlight.

At these turns in our intellectual tradition, it is the imagination that seems to negotiate the disagreements in philosophy as it mediates the binary ways of thinking that have continually emerged as points of contention in the history of Western thought: the *problem* of mind and body, the *disjunction* between form and content, the *conflict* between sensibility and understanding, the *tension* between diversity and unity, the *antagonism* between growth and collectivity. These disagreements continue to define large swaths of the philosophical landscape. Some new planting seems to be in order: the imagination clears a middle ground between parties and conceptions that have remained in longstanding opposition, a fertile ground where new forms of human flourishing can grow and be conceptualized.

I harbor a worry that has, to this point, remained buried in footnotes and offhand comments, namely that professional philosophy has become *intentionally* boring. The push to publish in competitive academic departments has encouraged scholars to take up ever-narrower philosophical questions. As their questions have "thinned out," the motivation behind their research has grown ever more questionable. The focus on the minutiae of certain fields of philosophy causes us to lose sight of the fact that philosophy, at its best, concerns the creative and imaginative business of living. I believe that a "focus" on the concept of the imagination has the opposite effect of this type of intellectual myopia. I hope that such a focus on the imagination might open up the possibilities of philosophy. I mean to encourage philosophy to take up human creativity not only as its topic of study but also as its guiding *telos*. Dewey suggested that philosophical disagreements—the bickering that results from comfortable boredom—reflects analogous conflicts that exist in the field of contemporary culture. Once again—and this time unfortunately—he seems to be on target. Squabbling in the fields of epistemology and metaphysics echoes the general dis-ease of a people who have grown weary with their playthings. Indeed, modernity's tendency to produce more distractions, both conceptual and social, has done little to relieve the symptoms of summer afternoon boredom. Have we reached the weary point where even the exciting and the obscene is overdone and passé? Do we live anesthetically—asleep to one another and to the potentialities that quietly underpin our lives? Has the gap between private impulse and public concern widened to a disturbing degree? Has our sense of a creative future been trumped by cynicism?

In the wake of these questions, which must be asked, the encouragement to "be imaginative" seems not just appropriate but frighteningly necessary. The encouragement does not supply new toys but a new way of engaging with them, a new way of feeling, a new way of combining ideas and things to compose a satisfying and unifying situation. My project begins to explain what it might *mean* to "be imaginative" and to highlight the role and significance of the imagination in the history of philosophy.

Attempts to situate the concept of the imagination in the history of philosophy have proved difficult. I will argue that they have been largely unsuccessful. Dewey recognizes this lack of success, commenting:

> "Imagination" shares with "beauty" the doubtful honor of being the chief theme in esthetic writings of enthusiastic ignorance. More perhaps than any other phrase of the human contribution, it has been treated as a special and self-contained faculty, differing from others in possession of mysterious potencies.[4]

Dewey makes several important points here regarding the traditional treatment of the imagination. First, the imagination is traditionally addressed by work in aesthetics and narrowly framed in aesthetic practice. According to this view, the imagination is associated almost exclusively with the creative arts rather than with epistemic processes. Imaginative work is the exclusive province of painters and children. The imagination is framed—like the canvases of Cezanne and Vermeer—on an easel, in a museum. Rational adults are expected to outgrow their imaginative bodies after a short and indulgent period of whimsical play.

As the discipline of aesthetics has become more of a self-contained field, a trend that began in the eighteenth century and continues to this day, the marginalization of the imagination has become more pronounced. Many scholars note, however, that an understanding of the imagination as a mere play of images distinct from the serious work of reason predates the rise of the discipline of aesthetics. Indeed, the understanding can be traced back to the Platonic dialogues or, more accurately, to the Platonism that arose from these works. The prejudice against the imagination and imitative arts more generally is typified by Socrates' insistence in the *Republic* that "the imitator [the image maker] has no knowledge worth mentioning."[5] The yawning gulf between reason and imagination opens further in the *Ion*: "[the] imitator is a flighty thing, a winged and flying thing; and he cannot make poetry and images until he becomes inspired and goes out of his senses and no reason is left in him."[6] Indeed, it is in light of the imagination's alleged lack of reason that Socrates suggests that imaginative poetry does not enliven but rather corrupts young minds. This Platonic prejudice has been taken up into "treatments of art and beauty in which the imagination is seen chiefly as a creative and wild artistic faculty unsuited for scientific or theoretical cognition."[7]

Early interpretations of the dialogues that address the place of the imagination often neglect the distinction between *eikasia* (the ability to disclose a type of novel continuity from observed cases) and *phantasia* (the ability to generate fanciful images). By the fourteenth century, most interpretations had wholly overlooked the character of *eikasia* and mistakenly equated *phantasia* with the imagination as a whole. On these grounds, commentators and modern thinkers marginalized the concept of imagination as a figment of fancy in their epistemological and metaphysical accounts.[8]

Dewey is aware of the imagination's tarnished history. He suggests that the prejudice against the imagination in Platonism has been accentuated by a type of faculty psychology in which the human mind is partitioned into various separate components that serve distinct functions. In this respect, the imagination has been rendered as an affective and emotional component of cognition, one sequestered from the logical, analytic, and rational centers of mental life. This hard line between bodily emotion and rationality can be proximately traced to eighteenth-century faculty psychology but ultimately stems from the modern division of the cognitive field found in Descartes' *Meditations* or his intellectual Christian precursors. According to Dewey and the pragmatists, this approach to the cognitive faculties and the emotional imagination in particular coincides with the mind-body dualism initiated by Descartes. That position continues to underpin much of Western thought.

Dewey was by no means the first to voice this criticism. In the opening decades of the nineteenth century, German transcendental philosophy made its way to America. American thinkers began to question the standard model of the imagination as *mere* fantasy as well as the presuppositions that support modern notions of thought and experience. For these thinkers, to reframe the imagination was not only to draw it out of the aesthetic sphere but also to broaden its scope in order to expose its epistemic and ethical import. One task of this book is to identify important moments in which the imagination emerges not only in the course of artistic life but also in the fields of epistemology, ontology, and ethics. Dewey seems to reflect a similar sentiment in his comment that "esthetic experience *is* imaginative. . . . [However,] this fact, in connection with a false idea of the nature of the imagination, has obscured the larger fact that all conscious experience has of necessity some degree of imaginative quality."[9]

When Dewey states that "all conscious experience" bears an imaginative quality, he is making the radical suggestion that all human affairs, including the affairs of mathematics, logic, and science, are shot through with imagination. This is a suggestion that I will explore in some detail. Imagination functions not just in the discovery of new ideas but also in the very logic and dynamism of reasoning. While Dewey suggests the importance of the imagination in thought, he is unable to provide a satisfying argument that might support this hypothesis. In works such as *Experience and Nature* and *Art as Experience*, Dewey develops a phenomenology of imagination, a way that we might experience imaginative events. But he fails to develop an epistemology that is both analytically rigorous and thoroughly imaginative. It is through Dewey's work that many contemporary pragmatists have come to think about the importance of the imagination in American pragmatism, in the history of philosophy, and more generally in the life of thought. The current project seeks nourishment elsewhere, finding imagination in other plots of the philosophical garden, as initially sown by the German Enlightenment thinkers of the eighteenth century and tended in the late nineteenth century by the American pragmatist Charles Sanders Peirce.

Unearthing and Replanting the Imagination

> It remains true that there is, after all, nothing but imagination that can ever supply [one] an inkling of the truth. He can stare stupidly at phenomena; but in the absence of imagination they will not connect themselves together in any rational way.
>
> Charles Sanders Peirce (1898)

Whenever human beings think, they think through the imagination.[10] But if human beings do think through imagination, what exactly are they thinking through? Dewey rightly notes that the imagination has long been regarded as a faculty "differing from others in possession of mysterious potencies." The term "the imagination" has been used as a placeholder for processes and events that defy explanation. For this reason the imagination, almost by definition, remains an enigma. It seems both impossible and unwise to "clear up" this mystery, to set things straight, to explain the imagination on no uncertain terms. A better course is to explore the conceptual space that "the imagination" has occupied and to see what we find.

This is not a project *on* aesthetics as it is traditionally conceived. This is rather an attempt to explore the way in which human cognition is grounded in and motivated by a process that has been treated and downplayed as *merely* aesthetic. This exploration opens new ways of thinking through the spontaneity and adaptation of human thought, the embodiment of cognition, and the continuity between thought and nature on the whole. Similar explorations exist in the Western philosophical canon and have met with varying degrees of success, explorations that have helped clear a path for an investigation of the imagination. Two historical interlocutors—Immanuel Kant and Charles Sanders Peirce—have helped frame this argument. Their treatment of the processes of imagination points to its philosophic significance. Likewise, current investigations into the nature of human cognition in cognitive neuroscience and empirical psychology echo the suspicions of these philosophical figures and point to meaningful paths in any future investigation of the imagination.

Recognition of the imagination's centrality to thought emerged at a poignant moment during the European Enlightenment. Indeed, its expression in the later work of Immanuel Kant signaled a radical turn in Western thought, which had to that point downplayed the imagination's significance. Perhaps "enlightenment" had simply become too boring. Perhaps the promises of the Enlightenment, the triumphant promise of reason and freedom, had fallen flat. Perhaps the promise itself had shown itself to be ill-conceived. In any event, Kant's rendering of the imagination and the schematism in the *Critique of Pure Reason* stands as an interesting moment in which imagination is considered, or at least acknowledged, as a significant force in human cognition. Prior to this point, the imagination was dismissed as a fanciful play of images, as a faculty unfit for the gravity of serious thought. Kant inherits this understanding of the imagination and struggles constantly with this inheritance. The struggle reveals itself in the first *Critique*, where his acknowledgment of imagination's importance is less than fully fledged. Kant remains ambivalent concerning this creative power—simultaneously attracted and repelled by its force. Like a child on a summer afternoon, Kant is not sure he wants to face the implications of the imaginative imperative. Kant works with the imagination, begins to draw out its character, but fails to develop fully the role of creative imagination in human thought.

In the first *Critique*, imagination functions as a schematizing process that mediates between sensibility, which Kant regards as passive and receptive, and understanding, which is described as active and legislative. Against the backdrop of this sensibility/understanding dualism, Kant never adequately explains how imagination can bridge the gap between the two realms of cognition, with one tied to bodily sense and the other bound to formal mental sensibility. The binary ways of understanding cognition that Kant establishes in the architectonic of the first *Critique* remain unmediated and pragmatically problematic. That is also to say that understanding and sensuous experience are continually torn asunder. In the *Critique of Pure Reason*, Kant is unable to envision the model, later developed by American pragmatists, in which human understanding is grounded in the field of human experience, embodied and sensuous. Additionally, he refuses to acknowledge, for various reasons, the way in which the imagination might serve an originating rather than mediating role in thought. His pragmatic turn had yet to come.

This turn comes in the *Critique of Judgment.* The imagination had not received its fuller treatment until Kant wrote this final critique in 1790 and the *Opus Postumum* in 1794. In these works, he describes the imaginative process in detail, highlighting its unique mediative character, underscoring its embodied nature in human genius, and indicating how it might serve a creative role in his epistemology. Kant suggests that genius embodies imagination, a "natural gift" (*ingenium*). Genius and the creative imagination that supports it serve as the bridge between nature and human freedom, spheres that had long been disjunct in modern philosophy. In the *Critique of Judgment*, Kant begins to draw out the strands of continuity among various processes of human thought, once again highlighting the role that imagination plays in establishing the continuity between nature and human freedom but also between subjective sensibility and objective understanding.

Despite these gestures, it is by no means clear how the imagination might fit into Kant's critical project on the whole. His work through the imagination remains ostensibly "aesthetic." His approach therefore effectively marginalizes the imaginative process as if it were not central to the broader fields of epistemology and ontology. Ultimately, Kant's treatment downplays the way in which imagination might transcend its traditional rendering as merely subjective or emotive. Notwithstanding his comment in the First Introduction to the third *Critique* that the work represents the culmination and mediation of the first two *Critiques*, it remains open to debate how this could be the case. In the third *Critique*, therefore, while Kant provides a phenomenologically rich account of the imagination—its structure, movement, and originality—he fails to incorporate this account into his earlier work on epistemology and ontology. Such an integration is, however, of vital importance if we are to understand the central place of the imagination in human thought. Such a project requires reframing the ostensibly aesthetic rendering of the imagination to draw out its epistemological and ultimately ontological implications.

Peirce's Imaginative Inheritance

The seeds of an adequate theory of the imagination may have been sown by the German idealists. But I will argue that they germinated and grew in a nurturing American soil. They came to life, rather quietly, in inquiry, in logic, and in ontology in the work of Charles Sanders Peirce. I will draw heavily on Peirce's epistemology and metaphysics to argue that his work provides a way to round out and deepen the description of the creative process initiated by Kant.

It is widely acknowledged that American pragmatism—most notably the work of John Dewey and William James—inherited an aesthetic sensibility that can be traced to European figures such as Henri Bergson and Friedrich Schiller. This sensibility has been repeatedly expressed in the secondary literature. Nevertheless, scholars still fail to take seriously the claim that human cognition, *on the whole*, thinks through the imagination. Often the imagination is still described as playing around on the outskirts of cognition, as not being central to human knowing. The study of the imagination continues to be addressed as an aesthetic subset of American philosophy. In the case of Peirce, commentators with only a few notable exceptions overlook the way in which the process of the imagination influences his development of logic, epistemology, and ontology. Peirce himself often seems to ignore this influence, commenting, "as for aesthetics, although the first year of my study of philosophy was devoted to this branch exclusively, yet I have since so completely neglected it that I do not feel entitled to any confident opinions about it."[11]

That said, I believe Peirce's work serves as an appropriate touchstone in contemporary philosophy's rethinking of the imagination as well as philosophy's rethinking of itself. John Dewey reflects a similar belief in 1932:

> What professional philosophy most needs at the present time is a new and fresh imagination. Only new imagination is capable of getting away from traditional positions and schools. . . . Nothing much will happen in philosophy as long as the main object among philosophers is defense of some historical position. *I do not know of any other thinker more calculated as Peirce to give emancipation from the intellectual fortifications of the past and arouse fresh imagination.*[12]

Dewey urges us to the imaginative imperative, to explore the concept of the creative imagination. He indicates that Peirce's thought might provide resources for revitalizing the aesthetic sensibility, to replant it in a variety of fields of study. More recently, Douglas Anderson's work on Peirce has made a similar point, exposing the place of the imagination in Peirce's corpus. Anderson writes, "there is an implicit theory of artistic creativity in Peirce's system which needs to be brought out."[13]

My project follows Anderson's lead in a significant respect. It makes the case that the most puzzling concepts of Peirce's system can be best understood as dependent on the creative imagination. This project tracks and develops a specific lineage from Kantian aesthetics to Peirce's thought along the lines of Hookway's 1985 comment that Peirce might agree with large swaths of the *Critique of Judgment*.[14] This agreement is not coincidental. I argue that Peirce read, and agreed with, large swaths of Schiller's *Aesthetic Letters*, though without an appreciation of Schiller's debt to Kantian aesthetics.

Peirce's development of the concepts of abduction (hypothesis formation), musement, and synechism depends on the unique structure of imagination posited by Kant and other German idealists in the eighteenth century. Embodied inquiry, proposed by Peirce in the late 1860s, well before the work in psychology by James and Dewey, bears the mark of the imagination as well, as evidenced by the similarity between the embodied and situated character of the aesthetic genius and the pragmatic inquirer. The notion of "continuity," the illusive holy grail of Peirce scholars, can be best negotiated, if not adequately explained, with reference to the unique structure of artistic imagination.

All of these claims belie a suspicion that the analytic rigor of Peirce's work can be best read, despite his demurral vis-à-vis the discipline of aesthetics, as an extension of the synthetic and creative force of the imagination. While Kant describes the character of aesthetic imagination without exposing its connection to epistemology and ontology, Peirce develops an imaginative epistemology and ontology without giving the imagination *qua* imagination its due. Work is needed to expose Peirce's assumptions, an aesthetic inheritance critical to his entire philosophical project. This work seems important in the scope of contemporary Peirce scholarship to the extent that current work seeks to integrate "the aesthetic or metaphysical Peirce" with "the analytic or logical Peirce." This project aims to avoid and to counteract the tendency of commentators to describe Peirce's interests in either one or the other of the aforementioned ways. Peirce re-turns the concept of the imagination in just such a way that its epistemological significance can be seen. Along these lines, it will be shown that these moments of Peircean inquiry and logic are, at their core, imaginative. In the process, I will outline a meaningful relation between the philosophical fields of aesthetics, epistemology, and ontology that most American pragmatists, with varying degrees of success, have sought to elucidate.

Another important continuity needs to be stressed. I take a moment to discuss the historical approach of my investigation. By arguing historically and analogically, by way of relative similarity and difference, I flesh out another strand of continuity between the American tradition and an earlier period of the history of philosophy. I propose an interpretation of Peirce that seeks to highlight the effect that a particular strain of German idealism, the strain that respects the imagination, had on his lifework. In this respect, and in light of the common criticism that American philosophy is done without a genuine understanding of the Western canon, the project seems worthwhile in and of itself. Much more importantly, however, is the particular suggestion that this story makes in regard to the nature of the imagination, namely that an imaginative process characterized by growth and spontaneity underpins and motivates human conceptualization. It should, however, be noted that I do not provide an exhaustive account of the way in which Kant's corpus influenced Peirce's philosophy. Rather, I sketch out a line of thinking about the imagination that can be traced from Kant's critical project to Peirce's writings.

Kant suggests, and Peirce maintains, thinking through the imagination requires one to think through the *nature* of cognition *tout court*. In this sense of "thinking through," I mean that an investigation of the unique character of the imagination encourages them to think about the natural basis of creative thought, to rethink the relation between nature and human cognition. As mentioned earlier, Kant begins to present a fuller conception of this connection in the third *Critique*, in the sections on genius as the *ingenium* (the natural gift). In these sections, he continues to deepen the similarity, if not continuity, of the processes of imagination and organic life (*Leben*). Peirce makes more explicit this continuity in works such as "The Order of Nature" (1878), "The Law of Mind" (1892), and "Evolutionary Love" (1893). These essays, among others, represent Peirce's attempt to draw imaginative consciousness from the narrow realm of the *strictly* human, to stretch the imaginative process beyond the constraints established by Cartesian mentalism. Our contact with the natural world is not simply a matter of "knowing," "thinking," and "understanding." Indeed, imaginative thought is intimately *in touch* with the world in so far as it is an outgrowth of novel biological processes that obtain in our embodied and thoughtful lives.

Peirce overcomes the legacy of Cartesian dualism by exposing the natural basis of imaginative processes, or, more dramatically, by underscoring the way in which nature, in its structure and dynamics, is continuous with the movement of the creative human intellect and artifice. For Peirce, the imaginative inquirer is not *apart from* but rather *a part of* the natural world. This "thinking through" of the imagination sets the stage to reframe ontology and nature, a process that characterizes both Kant's later works and Peirce's naturalism. Here, once again, it is worth noting that we extend the concept of the imagination from the ostensibly aesthetic to the traditionally epistemological and on to the speculatively ontological.

As the argument of the project begins to address the ontological implications of thinking through the imagination, that is, thinking through the *nature* of human creativity and through the embodiment of thought, my approach becomes particularly susceptible to criticism. It is vulnerable insofar as the argument could be confused with either a type of biological determinism or a naïve panpsychism. The project could, on the one hand, be accused of the reduction of human imagination to a set of particular biological functions or, on the other, be seen as the suggestion that nature is the work of some undifferentiated and ideal Mind. I take care to avoid both of these accusations. In this regard, the selection of Peirce as an interlocutor who addresses the imaginative character of thought is critical to this project. Peirce, unlike his contemporaries Josiah Royce, F. H. Bradley, and Borden Bowne, has not been branded as sympathetic to the cause of idealism. Indeed, Peirce is more often than not regarded as the consummate scientist. However, unlike John Dewey and William James, in essays like "A Guess at the Riddle," Peirce became increasingly willing to venture guesses in order to unravel the riddle-like pattern of natural being. These guesses draw Peirce into the field of ontology. In his work after 1884, Peirce endeavors to expose the way in which scientific investigation leads to a reformulation of nature on the whole. In 1898, Peirce concludes from these highly rigorous investigations:

> But if there is any reality, then, so far as there is any reality, what reality consists in is this: that there is in the being of things something which corresponds to the process of reasoning, that the world *lives*, and *moves*, and HAS ITS BEING in a logic of events.[15]

If I have been successful in my analysis of Peirce's earlier work, I will be able to conclude that the logic of events to which Peirce refers can be most appropriately and most succinctly described as the pattern and process of the imagination. I will be able to suggest that "the world lives and moves and HAS ITS BEING" in large part in the creative imagination. This far-flung claim can be more modestly stated: I hope to show how organic processes—creative, flexible, and dynamic—are continuous with the human imagination. This seemingly idealistic suggestion will follow from a careful analysis of Peirce's sustained study of logic and epistemology. This analysis will expose the imagination as a process of the world rather than "merely" an ephemeral figment of subjective fancy. In the spirit of Peirce's interest in the integration of philosophy and the empirical sciences, I extend this suggestion via a sustained investigation of the cognitive sciences.

The Imagination and the Contemporary Empirical Sciences

Chapters 2 through 6 focus on the work of Kant, Schiller, and Peirce. In them, I attempt to trace the movement of the imagination from these philosophers' respective studies of aesthetics and epistemology to their speculations with regard to ontology. That is to say that these thinkers begin by identifying the way that the imagination is at play in human artistry and cognition and then ask what sort of ontological under-pinnings must be secured in order for this imaginative play to take place. In the final chapters of this book, I highlight a similar train of thought, and I use contemporary linguistics and the cognitive sciences to draw out more fully the nature of the imagination.

With this strategy in mind, chapter 7 opens with a description of metaphor and image schemas as developed in contemporary linguistics. Here we can identify resonances of both Kant and Peirce on the topic of the imagination. In the late 1970s and early 1980s, metaphor studies sug-gested the ubiquitous role that analogy and metaphor play in the cre-ation of human concepts and meaning. As George Lakoff and Mark Johnson have proposed, imaginative metaphors serve as the glue of human language and cognition.[16] These metaphors are characterized by their flexibility and novel uses. Metaphor, literally translated from the Greek as "transference" or "to carry across," allows us to develop novel expressions by mediating between (among?) various forms of human experience. Metaphors are ways of handling and generating novel occur-rences and symbolizations. I examine the novelty of metaphors in refer-ence to our earlier discussion of the imagination. This will constitute the epistemic/aesthetic side of our investigation; what is seen as most vital about human thinking is its creative and spontaneous character, a char-acter that has long been associated with the imagination. Recent meta-phor studies cut in another important direction in our study of the imagination by suggesting that the development of conceptual metaphor and meaningful analogies rely, almost exclusively, on the physical em-bodiment of human beings. This observation encourages us to dig down into the ontological foundations of the imagination, investigating its bodily and biological basis. A preview or promissory note is warranted here. Consider the following expressions: I am feeling *up* today; the stock market went *down*; I just couldn't *grasp* what she was saying; the concept *went over my head*. These expressions, which rely on a few central metaphors, find their meanings in the forms of human embodiment. Humans come by imaginative meaning only in their organic, spatial, and temporal relations with the world. The organism-environment coupling, long studied in the fields of biology, ecology, behavioral psychology, and phenomenology, makes possible the emergence of stable and meaningful metaphors. This coupling is imaginative in its unique dynamic and arises from specific patterns of organism-environment interaction. Or, more accurately, the study of various forms of metaphor exposes the way in which our language and thought may depend on our natural embodiment for its meaning. In short, metaphor studies indicate that our thinking might literally be a form of genius (*ingenium*)—a gift of nature.

In my attempt to develop the role of the imagination in epistemology and ultimately in nature, I take a second look at the organism-environment nexus in which imagination seems to reside. I draw on biology and cognitive neuroscience to investigate the imaginative character of human life. I do this not only to provide evidence that the imaginative process of metaphor has its basis in the biological organization of a given human organism and equally in subject matter afforded by environmental circumstances. I also hope to venture an even bolder claim: that *any* given organism has its vital basis in a force and a dynamic that is best described in terms that have been historically reserved for the concept of the imagination.

The first, and less controversial, of these two suggestions has been advanced by cognitive neuroscientists and empirical psychologists alike. Antonio Damasio puts it in the following manner:

> From an evolutionary perspective, the oldest decision-making device pertains to basic biological regulation; the next, to the personal and social realm; and the most recent, to a collection of abstract-symbolic operations under which we can find artistic and scientific reasoning . . . but although ages of evolution and dedicated neural

*systems may confer some independence to each of these reasoning/
decision-making "modules," I suspect they are all interdependent.
When we witness signs of imagination and creativity in contemporary
humans, we are probably witnessing the integrated operations of sun-
dry combinations of these devices.*[17]

The sensibility that Damasio expresses—that complex processes of the
human imagination have their basis in an interdependence and interaction of "lower" physical systems—has been repeatedly expressed by the
scientific community for more than a century. Indeed, it seems fitting to
recognize Peirce's comments in his 1879 "Thinking as Cerebration" as a
preliminary step in this direction. Recent studies, however, have been
able to expose explicit ways in which the processes of cerebration give
rise to effective metaphors and imaginative consciousness. These studies
include Laila Craighero's investigation of the mirror neuron system,
Timothy Rohrer's investigation of cross-modal neural mappings, and
Gerald Edelman's studies on neural dynamics and synaptic selection.
Donald Tucker's work on the architecture of the brain also helps explain
the development of imaginative consciousness. Consideration of the
projects of these scholars "fleshes out" the description of the imaginative
consciousness provided by Damasio and, to a lesser extent, the very pre-
scient Peirce.

Edelman's work proves particularly relevant to my argument. Not
only does it provide support for embodiment studies, but it also seeks
to describe imaginative consciousness without confining it to *human*
embodiment. The title of his recent book, *A Universe of Consciousness:
How Matter Becomes Imagination*, reflects the intent behind this claim,
namely that matter and the imagination are continuous.[18] I take only
slight issue with the way in which Edelman construes his objective. It
is not the case, I argue, that matter somehow miraculously "becomes"
imagination. Rather, the relations and dynamics of matter are already, in
a certain sense and quite amazingly, imaginative. This distinction aside,
Edelman's description of neural reentry and neural degeneracy resonates with early descriptions of the workings of the imagination found
in Kant and Peirce. His work seems to indicate that physical processes
themselves reflect the spontaneity, the continuity over time, and the
freedom within constraint that has been established as being distinctly

imaginative. Nature itself emerges in, and emerges as, imaginative/creative process.

Along these speculative lines, the final sections of this book sketch a complex process ontology that follows from our investigation of the imagination. This investigation, instead of locating the imagination in any *one* particular place, in any *one* particular work of art, in any *one* particular mental process, will, on the contrary, show the imagination at home in a variety of related events in the aesthetic, mental, and physical world. In so doing, the investigation demands a reconfiguration of imagination and reframes natural processes. Such a reframing is already underway in the field of complexity theory, in which scholars such as Stuart Kauffman and John Holland propose that workings of nature can best be described in the language of creativity, agency, and emergence. This is precisely the imagination language introduced in the opening sections of this book. For Kauffman, natural phenomena cannot be fully described—nor even partially described—by mechanistic, determinate constructs. Extending a Peircean claim, Kauffman suggests that natural being moves and lives in a logic of relations that has traditionally been reserved for the dynamics of human aesthetic imagination. The work of complexity theorists brings this suggestion to light and elaborates on the observations made by figures from the traditions of American pragmatism and German idealism.

The Boundaries of the Plot

Anyone who has spent any time gardening knows the disturbing feeling that creeps up in the late afternoon, the feeling of impending darkness. In the midst of cultivating a plot, time that once dragged on suddenly flies, and there is often a rush to finish the project before nightfall makes careful gardening impossible. In light of this fact, it seems best to choose an appropriately sized plot.

I have not taken this piece of advice. The plot that I have set out for philosophical investigation, which ranges over German idealism, American philosophy, and the empirical sciences, is one far too large to handle in a single book, much less a slim one like this. Here, I am reminded of Mary Warnock's words at the beginning of her treatment on the imagination:

"The imagination is a vast subject, and it may seem rash to treat of it in a relatively small book."[19] I follow Warnock in not apologizing for my overreach "too abjectly" since I do not claim to cover the subject of the imagination in any comprehensive way but merely to trace out certain aspects of the concept through three fields of thought. Other scholars, such as Eva Brann, have done impressive jobs in charting and surveying the *World of the Imagination*; I merely endeavor to cultivate and explore a very little chunk of it in the coming pages. In contrast to Brann, Warnock "traces a single thread through different accounts of the imagination [from Hume, to Kant, to Coleridge] . . . to see whether certain features of the imagination would emerge as essential and universal."[20] Likewise, I hope to show that the imagination serves as a point of meaningful continuity between three fields that are often tended separately. My project is similar in character and, indeed, could be regarded as an extension of Warnock's. It proceeds from her observation that "in Kant, it is the imagination which has emerged as that which enables us to go beyond the bare data of sensation, and to bridge the gap between mere sensation and intelligible thought."[21] While the next chapter underscores the "bridge" of the imagination in Kant's critical project, my intent is to get a bit more clarity about the nature, origin, and function of the bridge through Peirce and the empirical sciences. To this end, I could have turned to a variety of other thinkers that remain notably absent from the coming analysis, most notably Edward Casey, Maurice Merleau-Ponty, and Paul Ricoeur. Ricoeur's linguistic approach to the subject meshes very nicely with key insights provided by Johnson and Lakoff, addressed in chapter 7. Casey and Merleau-Ponty provide phenomenological accounts of the imagination that could supplement the description of Peirce's abductive logic and complement Dewey's analysis of aesthetic experience in *Art as Experience*. Merleau-Ponty's "Cezanne's Doubt" may be the most succinct continental rendering of Peirce's understanding of continuity presented in the "Law of Mind." I do not intend to argue these points here, for doing so would take us far afield and distract us from the project at hand.

What is the project at hand? It is to provide a provisional answer to the question: How do we think through the imagination? As mentioned earlier, an appropriate answer would be both phenomenological (in the spirit of Casey) but also empirical and evolutionary. Peirce is one of the few authors who was willing to take both of these approaches to the subject and attempt to unify them. In the end, Peirce's attempt fails, and fails for good reason, but the effort is instructive and tells us something significant about the nature of creativity. Given the number of times that Dewey has already been mentioned, one might be surprised to find that my treatment of the imagination avoids, for the most part, the growing literature on Dewey and aesthetic experience. Dewey and scholars of Dewey such as Thomas Alexander are exceptionally good at sketching out the phenomenology of aesthetic experience. When I can, therefore, I occasionally draw on their insights and observations. This being said, I believe that Peirce's philosophical project is more suited to thinking through the biological basis of the imagination, to investigating the natural preconditions of creative experience. Both Peirce and Dewey held that mind was embodied, structured by the bodily comportment of the human organism. Both, according to Anderson, maintained the belief in the "reality of habit and the reality of possibility."[22] I recognize the affinities between the two thinkers, but I understand Peirce as taking a slightly different approach to the study of the embodied mind's habits and possibilities. For Peirce, to say that the mind is embodied is to suggest that the workings of human cognition are structured not only by bodily comportment but also by emergent dynamics of the natural world at every level of analysis. It is for this reason that Peirce shifts from an analysis of physics, to chemistry, to biology, and to psychology when he is fleshing out the "Law of Mind" in 1891 and 1892. In Anderson's words, "Peirce openly claimed that nature is itself shot through with reason, with habits."[23] Dewey, however, often seems to shy away from such assertions, seemingly aware that they would mark him as a type of traditional metaphysician. Peirce, however, repeatedly takes this risk. And it is this risk—along with Peirce's turn of phrase that expresses this risk—that drew me to think through the imagination with him instead of Dewey.

While this book shares common ground with a number of philosophers who have devoted time and energy to exploring the concept of the imagination, it deviates in substantial ways from others who have handled the topic. For example, I do not endorse aspects of Jean-Paul Sartre's description of "the imaginary," which, according to Sartre, stands against perception and intellection. For Sartre, the imagination is the enabling condition of radical freedom, that cognitive function which makes human beings uniquely different from the world of cats and rocks and slime molds. In contrast, I am interested in exposing the continuity between human creativity and the emergent creative forces of the physical world. Like Peirce, the story that I would like to tell about the imagination also includes slime molds. I agree with Sartre that the freedom that the imagination affords is radical to the extent that it deviates from the listless free time of the everyday (think of the freedom of summer afternoons), but it is a freedom that is found in maintaining a unique orientation to the world that is not wholly separable from the workings of our imaginative minds. The imagination, in this sense, is not a power that allows us to escape the limits of facticity and immanence but is rather one that creates an opening for us to explore and converse with a world that is alive and meaningful—creative in its own right. The imagination allows us to act in spontaneous and creative ways, but it also, and perhaps more miraculously, allows us to be in lively touch with a realm—like a plot of zinnias and hostas—that we are often inclined to objectify as dead and dumb. And hostas and zinnias are neither dead nor dumb. Indeed, I hope to show that the particular way that the natural world "lives and moves and has its being" is continuous with processes that we have long reserved for our descriptions of the imagination.

TWO

ENLIGHTENING THOUGHT:
KANT AND THE IMAGINATION

Framing the Imagination in Western Thought

The imagination is notoriously difficult to define.[1] Indeed, this difficulty may explain the fact that prior to the Enlightenment there was no attempt to develop a unified theory of the imagination. In the history of ancient Greek philosophy, its amorphous character contributed to its being treated in two distinct, albeit related, ways. On the one hand, imagination was defined in terms of inspired artistic expression, outside the realm of explanation and description. On the other, it was described as a mysterious mental faculty that somehow accomplished the impossible, bridging the divide between the world of sensation and the world of thought.

Mark Johnson, in his *Body in the Mind*, directs us to a passage in the *Ion* where Plato advances the first interpretation of the imagination as a function of inspiration. Plato writes:

> The poet is a light and winged and holy thing, and there is no invention in him until he has been inspired and is out of his senses, and

{ 25 }

the mind is no longer in him: when he has not attained to this state, he
is powerless and is unable to utter his oracles. Many are the noble words
in which poets speak concerning the actions of men; but like yourself
when speaking about Homer, they do not speak of them by any rules
of art: they are simply inspired to utter that to which the Muse im-
pels them, and that only.[2]

According to Plato, the creative imagination only operates in moments
of ecstasy, moments in which a person stands outside (*ex-stasis*) herself,
possessed by the artistic spirit. The suggestion that the imaginative poet
is "out of her mind" led Plato to assert that the imagination necessarily
stood against reason. The poets, along with the sophists, merely produced images and appearances that led individuals away from truth.
This position is reaffirmed throughout book 10 of the *Republic*, where
Plato states that the poets remain "thrice removed from the truth." This
assertion suggests that they do little more than enflame the negative
emotions of the public. To say that poets do *little* more than play on the
emotions is, according to Plato, no small claim: the imagination of poets
has the power to convince the public that the appearances represented in
their art are in fact real. Plato thus maintains that the imagination as
creative fantasy is necessarily dangerous and should be banned from the
ideal state.

Aristotle opposes Plato in this regard, making a clearer distinction
between the imagination as a cognitive faculty and its role in creative-
artistic expression. In both these characterizations, Aristotle acknowledges imagination's vital importance. He devotes book 3 of *On the Soul*
to the explanation of imagination as an indispensible part of human
thought, noting its ability to mediate between abstract conception and
sensuous perception. He claims that the imagination is different from
both the sensation that initiates human thinking and higher forms of
cognition that grant the possibility of formal judgment. Aristotle, however, is clear to state that imagination is not found without sensation and
that judgment is not found without the imagination. It is in this formu-
lation that we first come to see the type of mediating role that the imagination will take on in the Enlightenment tradition of the eighteenth
century. Additionally, Aristotle provides an alternative to Plato's rendering of the imagination as a creative impulse that should be tamed by or

exiled from the realm of cognition. In the *Poetics*, he acknowledges that the imagination can be used irresponsibly by poets or playwrights but suggests that, despite this danger, the imagination is a necessary force in creating a meaningful human life. According to Aristotle, *catharsis*, the outpouring of emotion prompted by the imaginative drama of a playwright and actor, serves as a healthy outlet for members of a given audience. Catharsis creates a type of escape valve for the public, who can deal with intense emotions in the context of an imagined situation rather than in the high-stakes business of daily life. The imagination of the poet serves another important function, Aristotle explains, by inviting others into a fictional world that is realistically felt. These imagined worlds serve a pedagogical function by allowing an audience to experience or sympathize with a variety of morally significant scenarios. While these scenarios have to be carefully constructed, the play of the imagination is the only way for members of an audience to gain sympathetic access to moral situations and to learn these situations by heart.

This brief survey of the archaic understanding of the imagination will serve as a helpful frame for our discussion of the imagination in the writings of Immanuel Kant. Kant was very much aware of the imagination's tarnished history in the Platonic tradition and of the attempts made by Aristotle to recuperate the concept. Here we find a thinker who is willing and able to work through the imagination, identify the danger associated with it, and develop a formal theory of cognition that begins to recognize its importance.

Kantian Critique: Sowing the Seeds of the Imagination

What role does the imagination play in Kant's critical project (1770–1804)? With regard to "thinking through the imagination," the answer to this question accomplishes three goals: (1) It exposes an opening in Kant's earlier works—a type of cognitive disunity between rationality and bodily sensibility. This "gap" in thinking characterized modern thought on the whole and called for a type of mediation. The later Kant, and the movement of American pragmatism, can be understood as an attempt to bridge this gap.[3] (2) It highlights the way in which Kant in the *Critique of Pure Reason* begins to answer this call for mediation with his development of the imagination, schematism, and the unity of apperception. Kant's answer at this point, however, is not fully fledged. He does not elaborate it until the *Critique of Judgment*. There, he acknowledges the way in which the divided architectonic of the first *Critique* gave rise to his interest in aspects of imagination such as reflective judgment, genius, and aesthetic common taste. Here, we catch a first glimpse of the imagination's distinctive character.[4] (3) Finally, it examines the turn in Kant's thinking when this imaginative ordering is brought to light. Recognition of the vital role of the imagination in thinking involves a revision not only of Kant's epistemology but also of his ontology and metaphysics. For Kant, this recognition entails a movement away from the dualistic structure of the first *Critique* and toward a pragmatic rendering of human inquiry. Additionally, in this turn to the imagination, he begins to outline what would later be taken up by Peirce and other American pragmatists as an agent-based ontology. It is in this sense not surprising that Kant's detailed investigation of the imagination reaches maturation in the *Critique of Judgment* and terminates in his *Opus Postumum* and his *Anthropology from a Pragmatic Point of View*.[5]

It is in these pragmatic accounts that Kant provides the most straightforward depiction of the imagination—now described as a "natural gift"—as being both inventive-productive and recollective-reproductive. This twofold character of the imagination underpins Kant's epistemic claims as well as his later theories of being. As the imagination moves to the foreground, nature, often presented in the first *Critique* as detached and unavailable to human thought, becomes alive, purposive, and immediately present in the thinking of imaginative genius.[6] This transition is especially important in light of a similar yet more thorough transition made in the work of C. S. Peirce.

The Problem and Imaginative Solution of the Critique of Pure Reason

When Kant completed the *Critique of Pure Reason* in 1781, he presented it as a response to the presiding epistemological position of the period, which was most poignantly reflected in the empiricists of the English school. The *Critique* is not a criticism or attack on pure reason but rather a critical analysis that sets the limits and exposes the possibility of objective knowledge of the natural world. This objective knowledge is not derived from the senses but rather from knowledge produced by the inherent structure and nature of mind. In this sense, Kant's project throws down a gauntlet at the feet of John Locke, who suggests that all knowledge arises from the senses, and of David Hume, who suggests that epistemic certainties are but probabilities derived from distinct sensations. Kant realizes that necessity and certainty cannot be supplied by what Plato called the "rabble of the sense," and in light of this fact, he asks if there is not an empirical knowledge whose truth is available to us before experience, that is available to us in understanding "a priori." This question is reiterated in his intention to examine the possibility of making synthetic a priori judgments and elaborated at length in the following passage:

> Experience is by no means the only field to which our understanding can be confined. Experience tells us what is, but not that it must be necessarily what it is and not otherwise. It therefore never gives us any really general truths; and our reason, which is particularly anxious for that class of knowledge, is roused by it rather than satisfied. General truths, which at the same time bear the character of an inward necessity, must be independent of experience—clear and certain in themselves.[7]

Here the distinction between sensible experience and understanding emerges. Sensation initiates the project of thinking but does not supply the "really general truths" that thinking requires. The organizing function of the understanding must operate beyond the "confines" of sensibility. As Kuno Fischer notes in an early commentary (1866), "sensibility and understanding are cognitive faculties that differ not in degree, but in kind. . . . This determination of the distinction between sense and understanding is the first position taken by critical philosophy."[8]

In the preface to the first *Critique*, Kant restates this distinction by means of a situating query: "My question is, what can we hope to achieve with reason, when all the material and assistance of experience is taken away?"[9] On Kant's terms, understanding can achieve nothing if it cannot transform the chaotic manifold of sense, in all of its unorganized stimuli, into ordered knowledge. This transformation occurs on two levels.

First, the manifold of unordered sense is ordered by the forms of intuition—space and time. These forms are given a priori in the sense that they are not derived from experience but rather create the conditions for the possibility of ordered sense. It is worth noting that this process does not happen automatically; perception is an active process by which the mind uses the forms of space and time to gather the unruly stimuli of the world. This stage of transformation is addressed in detail in the section on the Transcendental Aesthetic. Kant is using the term "aesthetic" in its original sense of pertaining to the feelings and sensation.

The second stage of transformation occurs when these developed perceptions are synthesized under the forms of conception, the categories of cognition. For Kant, the category operates as a kind of lawgiver who controls the unruly mob of the senses. This move is described in detail in the section on the Transcendental Logic. Here we can glimpse, although we do not yet fully grasp, the extent to which Kant wishes to separate the realms of sense and feeling (the aesthetic) with the sphere of understanding (the logic).

In the first subsection on the Transcendental Logic, Kant presents understanding as the source of certain pure concepts that are known a priori and are the conditions for the possibility of any meaningful experience whatsoever. In addition to pure concepts, Kant recognizes that there must also be empirical concepts that grant the recognition of particular objects in the world. Just as ordered perceptions involve an arrangement of sensations around the objects of space and time, conceptions involve an ordering of perceptions around the categories (in terms of various forms of quantity, quality, relation and modality).

In the section on the Transcendental Logic, it becomes clear that cognition does not begin *in* sensible experience but in the working of the categories "that have their seat in pure understanding." At another point, Kant writes that the categories "trace their origin to the understanding" and more specifically to the functions of judgment that "specify the understanding completely and yield an exhaustive inventory of its powers."[10] Insofar as the mind's function is the coordination of random experience, its coordinating structures stand as the substrate or, more accurately, the character of mind. In a language more suited to Kant, the categories "are the functions of judgment, insofar as they are employed in the determination of a given intuition."[11] In the "Transcendental Deduction of the Categories," Kant argues that all twelve of his concepts apply universally and necessarily to the objects that we intuit in our experience. Kant is not, however, out of the woods yet. While he may have proved the categories abstractly, he still has not specified the way in which these categories are applied to the objects of experience. In other words, he has not proved how his concepts could be "objectively valid." Allison puts Kant's crisis in other words: "under no circumstances ... can it be claimed that [the Deduction] succeeds in showing that the categories make experience possible."[12]

Now this is a real difficulty. Kant himself comments on this conundrum: "a difficulty manifests itself here that we did not encounter in the field of sensibility, namely how subjective conditions of thinking should have objective validity, i.e. yield conditions of the possibility of all cognition of objects."[13] This problem deepens throughout the Transcendental Deduction section as it becomes clear that Kant's categories of pure understanding remain pure and objectively valid only to the extent they are not derived from the cluttered manifold of sensibility. This contamination, however, seems unavoidable. As Kant admits earlier in the *Critique*, "all knowledge proceeds from sense."[14] Kant recognizes the aporia that he has encountered. While all knowledge might proceed *from* sense, Kant refuses to suggest that all knowledge arises *in* sense. This difficulty is poignantly expressed in "On Understanding's Relation to Objects as Such."[15] It is interesting to note that Peirce, more than a century later, comments on Kant's crisis, implying "that Kant draws too hard a line between observation and ratiocination."[16] Peirce's notes on the *Critique of Pure Reason* point to the fact that the line between ratiocination and observation reflects a more basic dualism in Kant's rendering of human knowledge. In his unpublished note, Peirce writes that "Kant says that while all our knowledge begins with experience, a part does not originate from experience."[17] In light of this tension, Peirce asserts that Kant "is a sort of idealist himself"[18] and is bound to maintain this disjunction (between observation and ratiocination) as a type of logical contradiction. Kant "falls into the habit of thinking that ratiocination only begins after [observation] is complete; and wholly fails to see that the simplest syllogistic conclusion can only be drawn by observing the relations of the terms in the premises and conclusion."[19] For Kant, pure concepts are unequivocally *not* empirical. Empirical observations are *not* pure.

As the problem between understanding and sensibility shows itself, Kant begins to formulate a cognitive process to mediate these two radically different modes of cognition.[20] Indeed, Kant insists that his transcendental philosophy is unique in its attempt to wed pure conception with the empirical instances of human experience. He writes that not only should his philosophy furnish the rules of conception. It should also identify "*a priori* the instances to which these rules apply."[21] To give up on this project would be to admit that pure concepts are mere logical forms devoid of all empirical content.

At the beginning of the "Transcendental Analytic," Kant asks two questions: "How can perception be subsumed under a pure conception? How can a category be applied in determination of an object of sense?"[22] In a certain sense, Kant is repeating himself, reopening the question concerning synthetic a priori judgments that initiates the work as a whole. At this point, however, Kant provides at least a temporary answer.

> Manifestly, there must be a third thing, which is homogeneous on the one hand with the category, and on the other hand with the object of sense, and which thus makes the application of one to the other possible. This mediating idea must be pure, or free from any empirical element, yet it must be at once intellectual and sensuous. Such an idea is the transcendental schema.[23]

With this, he offers us a vision that appears almost pragmatic. It reflects an odd departure from the form-matter distinction that marks most of the Kantian corpus. In the "Schematism" chapter, he promises his reader to outline the conditions by which an object of the world can be given "*in concreto*" but also in accord with the categories.[24] This amounts to a type of promissory note, recommitting Kant to the project of unifying the felt sensation and pure conception.

At many points in this section, Kant comments that the schemata are products of the imagination, the third and final faculty of cognition. The schema, however, is no mere product in the sense of being a thing. Rather, it is a continuous "unity in the general determination of sensibility." This distinction seems necessary. For while Kant occasionally describes the schema as a product of an imaginative process, he often describes it *as the procedure of cognitive ordering*. Imagination produces the schemata only to the extent that the schema *acts* in its function of mediating between pure, empirical concepts and objects of the world. Schema is rendered as both product and process. "The schema of sensible concepts . . . is a *product* as it were, a monogram of pure *a priori* imagination, *through which and in accordance with which*, images first become possible. These images can be connected with the concept only by means of the schema to which they belong."[25]

Allison helps unpack this passage by noting that the schema of an empirical concept can be regarded as the essential or generic features of the object falling under the concept, a sort of "monogram" or sketch of their defining features. This monogram, however, is operational—it is a monogram "through which and in accordance with which" representations are made possible. Like the art of tracing a monogram, the imaginative schema furnishes the rules by which an orderly process can be executed. The schema is also the product of this process and stands as a type of finished monogram. The schema is distinguished from the concept itself, which provides the rule for the schema's construction, and from the image itself of the particular object that falls under the concept. Only through this distinction and their ability to guide the production of images in space and time do schemas explain the applicability of the categories, or pure concepts, to appearances.

The role that schemata play in Kant's attempt to describe the synthetic character of human thought echoes an earlier suggestion that "synthesis in general, as we shall hereafter see is the mere result of the imagination, a blind but indispensable *art* of the soul, without which we should have no knowledge whatever, but of which we are scarcely ever conscious."[26] The irony of the imagination poignantly shows itself: the imagination is used in all thinking but is extremely difficult to define in thought. A monogram, regarded as the process of ordered tracing, cannot be grasped as one would understand a given object but is only known in the process of tracing. Similarly, the schemata, when regarded as a means by which ordered images appear, regarded as a collection of creative instruments, can only be understood in the midst of their use. The schemata serve as the bridge between the formal realm of Kantian understanding and the sphere of empirical perception. This imaginative bridging grants the possibility of all empirical thinking and all judgments concerning the state of the natural world. In conclusion, Kant writes that "without schemata, therefore, the categories are only functions of the understanding for producing concepts, but they present no object."[27]

The bivalent character of the schemata and the imagination—a character that allows imagination to function between the realms of sense and understanding—recalls Kant's framing of these cognitive realms in the opening of the *Critique of Pure Reason*. At that point, he ambiguously wrote, "Our knowledge springs from two fundamental sources of the mind which perhaps spring from a common, *but to us unknown root*."[28] In returning to the "blind" faculty of the imagination of which we are "scarcely ever conscious," does Kant return to the common root of understanding and sensibility? Kant does recognize the ambiguity that he has introduced with the suggestion that there may be a common mediating root of cognition, that there is a real difficulty in precisely defining the nature of the imagination. "This schematism of understanding, in its application to appearances and their mere form, is an act concealed in the depth of the human soul, whose real mode of activity nature is hardly likely ever to allow us to discover and have open to our gaze."[29] In the later sections of the first *Critique*, Kant retreats from his description of the "concealed art" of schematism, returning to the subject of the categories. Commentators have expended serious effort to show that the final sections really deal with the imagination. It seems more likely that Kant exposes the importance of the imagination yet remains hesitant to thematize the point. Kant glimpses the central role of the imagination in human thinking, but he pulls back from fully embracing this insight because of his commitment to the rigid dichotomies established in the first *Critique*.

This hesitation appears in Kant's tendency to subordinate the imagination to the understanding in the first *Critique*. While the imagination is crucial to synthetic understanding, it still serves the understanding in its synthesizing role. It serves understanding as a vassal who brings the wild mob of appearances under control. It is in this limited capacity that the imagination and its schematizing function operate in strictly productive and reproductive roles in the first *Critique*. As Mark Johnson points out, the imagination orders our images of objects in its reproductive function and forms the temporal unity of our consciousness in its productive function.[30] Kant had yet to develop a model of the imagination that acknowledged the principles of growth and spontaneity. That is to say, he had yet to make the imagination creative.

The reading of the first *Critique* as focusing on imagination has been advanced by various commentators throughout the twentieth century. These scholars, continental and American alike, highlight the constitutive character of the imagination in Kant's early thinking. In 1927, Martin Heidegger gave a series of lectures on the first *Critique*, which have been published in English as *The Phenomenological Interpretation of the Critique of Pure Reason*.[31] In the section entitled "Sensibility and Understanding: The Two Roots of Human Knowledge; the Common Origin of Both Roots," Heidegger addresses imagination as the common source of human cognition, the ground from which sensible determination springs. Heidegger goes so far as to suggest that, for Kant, imagination comes before (*vor*) understanding and sense. It is in this "coming before" that imagination serves as the synthesizing force of the manifold.[32] Most significantly, he makes the point that the imagination grounds all forms of judgment rather than particular forms of reflective judgment.

Additionally, Heidegger unpacks Kant's description of the schematism and the unity of apperception, highlighting the way in which the imaginative subject is both receptive and spontaneously active. Again, the imagination is rendered in its twofold character—at once reproductive and productive. The imagination is drawn out as an embodied and ecstatic process. Heidegger concludes his lectures with a section entitled "The Significance of Kant's Doctrine of the Schematism."[33] In it, he suggests that the emergence of the schematism was modernity's most meaningful flirtation with the imagination and that it forced Kant to revise his notion of judgment in the second and third *Critiques*. This suggestion is oddly similar to Peirce's comment on the schematism when he asserts that the discovery of the schematism ought to have commanded Kant's interest more thoroughly. Peirce writes that the "doctrine of the schemata can only have been an afterthought, an addition to his system after it was substantially complete. For if the schemata had been considered early enough, they would have overgrown his whole work."[34]

Just as Peirce called for a reappraisal of Kant's aims, Heidegger's lectures set the stage for a continental rereading of the opening moments in Kant's critical project. Hannah Arendt and Hans-Georg Gadamer both adopt Kant's notion of reflective judgment to motivate their respective investigations of epistemology and political philosophy.[35] John Sallis's *The Gathering of Reason* repeats, almost verbatim, Heidegger's reflections on Kantian imagination as provided in the first *Critique*.[36] Sallis attempts to argue that the Transcendental Dialectic, despite its seldom acknowledging this dependency, must rest on and be moved by the force of the imagination. The dialectic addresses apperception in detail as "the highest principle of human knowledge." Sallis writes that this unity of apperception "requires the imaginative synthesis and is dependent upon it."[37] Heidegger and Sallis encourage a reinterpretation of the *Critique of Pure Reason* but also point to the importance of the *Critique of Judgment* and Kant's later works. It is in these later works that Kant recognizes the ostensibly aesthetic process at the core of human cognition.

Along these lines, scholars such as Paul Guyer,[38] Henry Allison,[39] Mark Johnson,[40] Rudolf Makkreel,[41] John Zammito,[42] and Eckart Förster underscore the centrality of the imagination in Kant's critical period. This is also to say that they recognize the third *Critique* as the culminating and most interesting moment of this period. Allison provides a sustained analysis of the Transcendental Deduction of all three *Critiques*; he highlights the cognitive function of the imagination, emphasizing that the unity that the imagination imbues sensibility "precedes all concepts."[43] In *The Body in the Mind*, Johnson brings to light the tension in the first *Critique*—the way in which Kant separates the realm of empirical sense and pure cognition.[44] Subsequently, Kant dedicates an enormous amount of time and effort to the reconstruction of this epistemological Humpty Dumpty. Once the cognitive faculties of sense and conception are broken into pieces, however, they prove extremely difficult to put together again. Kant's dedication to the epistemological dualism of the modern age dies hard, precluding the full development of the imagination as a mediating process between these distinct cognitive realms. Kant merely points to the richer account of the imagination that Johnson begins to develop as being both constitutive and creative in human conceptualization and reasoning. I undertake a similar goal, extended along pragmatic lines. Förster's recent work on the *Opus Postumum* draws attention to the "gap" that Johnson and others identified in Kant's thinking. It was a desire to fill this gap that led Kant into his work after 1888, into a rethinking of the "*Selbstsetzung* [self-positing] that provides the schema for outer sense."[45]

The Limits of Knowing. Kant's Conception of the Noumena

To this point, our discussion has been occupied with the Transcendental Aesthetic and the Transcendental Logic. It has been concerned with the positive project of critique in which Kant sets out the conditions for the possibility of knowledge that is both synthetic and universally valid. We have seen in Kant's system that a gap exists between the manifold of sense and the unity of understanding, a gap that only the processes of the imagination can fill. I have suggested that in his first *Critique* Kant fails to draw out the creative principle of the imagination in full, placing it under the commanding hand of the understanding. While there may be shortcomings in the workings of the positive portion of the *Critique*, there are limits to Kant's model of cognition that he himself came to recognize.

The conditions that render cognition possible also confine it to a limited province. These limits are set out in the section on the Transcendental Dialectic. This section, in its development of the phenomena-noumena distinction, has earned Kant the questionable distinction of being a type of idealist. Peirce remarks on this fact, suggesting that instead of diving into the idealistic waters of the noumena, Kant would have been better served by a fuller development of the schematism. I want to examine briefly how Kant's assumptions led him to posit a phenomenal versus noumenal distinction that cannot be bridged. Peirce and his contemporaries avoid such a dichotomy by focusing on the primacy of the schematizing function of the imagination.

While Kant objects, and objects strongly, to the empiricists' belief that knowledge is reducible to a matter of sense, he does concede that there is no knowledge without initial sensation. Empirically meaningful thought proceeds from sense. This positive statement has a negative counterpart, namely that there is no knowledge without sense. More simply, no empirical cognition of objects is possible that does not first appear in sensible intuition. Cognition takes flight from the appearance of objects. Here, the loggerhead that Kant confronts forces him make the distinction between phenomena and noumena.[46] This distinction, in turn, encourages Kant to develop the concept of the noumenal "thing-in-itself" as the unknowable *ground* of human cognition.[47] This is also to say that, for Kant, the imagination has no access to the things of the world. The imagination provides the possibility of knowledge in the phenomenal realm by way of its synthesizing role, but it has no access to the things themselves that stand behind sensation. As Eckart Förster and others have noted, the gap that opens up between the noumenal and phenomenal and between the thing-in-itself and knowledge is never effectively bridged in the first *Critique*.[48] This fact will be decisive in the development of Peirce's epistemology, which eschews any notion of the thing-in-itself. While Kant forbids the imagination from apprehending things-in-themselves, Peirce's epistemological realism depends on a continuity between the imaginative process of the mind and the things of the world.

The Critique of Judgment: *Kant's Pragmatic Turn*

In his analysis of Kant's aesthetics, John Zammito suggests that the desire to bridge the gap between sensibility and understanding opens Kant's discussion of judgment, and more particularly reflective judgment, in the *Critique of Judgment*.[49] In the third *Critique*, in his attempt to unify sensibility and a type of communicable rationality, Kant enlarges his conception of judgment to include the imaginative processes of reflective judgment.[50] Kant introduces a new aspect of judgment that is distinct from his project in the first *Critique*, making room for a type of judgment that depends on "the free play of the imagination" and an elaboration and radical extension of the role imagination plays in the schematism. On these grounds, Eva Schaper notes that the schematism of the first *Critique*, which has posed problems for interpreters, may be elucidated by the *Critique of Judgment*. Schaper responds to befuddled interpreters by noting that "the Third Critique was still to come. Might it not shed some light on the chapter in which Kant speaks of the schematism as an 'art concealed in the depth of the human soul' (B 182)?"[51] Even in his later writings Kant continues to struggle with the possibility of synthetic judgments and begins to envision a more robust model of the imagination in the process. The "art concealed in the depth of the human soul" comes to the surface of the third *Critique*, and its full import is seen. I take Schaper's suggestion seriously that "aesthetic judgments of taste as they are discussed in the first part of the *Critique of Judgment* can be seen as paradigmatically exhibiting the ground for the possibility of judgment *tout court*."[52]

In the *Critique of Pure Reason* Kant seems to have envisioned only determinate judgment. But in the *Critique of Judgment* he comes to recognize a more creative and expansive dimension of judgment. The first *Critique* concerns itself with the possibility and limitations of judgments of nature that are furnished by the concepts of the understanding and could be considered universally valid. By contrast, the *Critique of Judgment* focuses on the possibility of judgment that appears to be rooted in a type of feeling that, despite this sensuous element, can still be universally communicated. Kant divides this unique type of judgment into three categories. First, he suggests that there are teleological judgments that are reflective. These are judgments made in regard to the purposes and possible ends of nature. Second, there are material judgments that I would describe as matters of personal preference. These judgments, while not matters of sheer fancy, cannot satisfy Kant's desire to hit upon a form of judgment that can be universalized. Finally, and most importantly for our purposes, Kant insists that there is a final category of reflective judgment that he terms aesthetic judgment. These are judgments of taste that, while possessing an important sensible component, have an intellectual and formal basis and therefore can be universalized. It is obvious that the imagination comes alive in the discussion of the play of reflective judgment. It becomes clear that Kant is developing an alternative to the determinate judgments of the *Critique of Pure Reason*.

I have suggested that the imagination and the schematism are quickly subdued by the limits of the understanding in the first *Critique*, constrained to separate "reproductive" and "productive" roles. Gasché echoes this point, describing the imagination as rendered in the *Critique of Pure Reason.* "In its ordinary employment, the imagination is not free at all. It operates under the strict rules of the understanding. It has no identity of its own, so to speak in its empirical employment."[53] In section 22 of the *Critique of Judgment,* on the other hand, the imagination frees itself from its determinant function and plays a *creative,* reflective role in aesthetic judgments.

In the first *Critique,* Kant writes that if the universal (the rule), or principle, or law is *given* so that the particular is subsumed under it, then the judgment is determinant.

> The determinant judgment only subsumes under universal transcendental laws given by the Understanding; the law is marked out for it, a priori, and it has therefore no need to seek a law for itself in order to be able to subordinate the particular in nature to the universal.[54]

In other cases, however, the forms of nature are so manifold, and there "are so many modifications of the universal transcendental natural concepts left undetermined by the laws given, a priori . . . that there must be laws for these forms also."[55] Kant in effect states that the forms of nature overrun the determinate concepts of the understanding described in the *Critique of Pure Reason* and must be brought under the universal in another fashion. This realization prompts a search for new "laws" of judgment and draws Kant to develop the character of reflective judgment and aesthetic ideas.

As opposed to determinant judgment, reflective judgment stands as "the capacity for reflecting on a given representation according to a certain principle, to produce a *possible* concept."[56] In a determinate judgment, the imagination synthesizes a sensible manifold under the guidance of concepts provided by the understanding. In reflective judgment, however, imagination plays with various possible syntheses, yet without the constraining role of any particular concept. In this respect, the imagination operates in a capacity of free play. Instead of operating in its determinate employment, whereby it functions as a particular proportion between an array of sensible intuitions and a *given* concept, the imagination now operates by an activity of reflection in which it harmonizes with a particular range of possible concepts of the understanding, without any particular determinate concept controlling the synthesis. Kant clarifies the initial distinction between reflective and determinant judgment in the Second Introduction of the *Critique of Judgment*. He restates this clarification in the *Logic* (1800).

> If the universal (the rule, the principle, the law) be given, the judgment which subsumes the particular under it (even if, as transcendental judgment, it furnishes, a priori, the conditions in conformity with which that subsumption under the universal is possible) is *determinant*. But if only the particular be given for which the universal has to be found, the judgment is merely *reflective*.[57]

Kant's comment that such judgment is *merely* reflective does not, despite our common understanding of "mere," diminish the importance of this cognitive faculty. Heidegger notes in his aforementioned lectures that Kant's use of *bloss*, often translated as "mere," can also mean "only," "simply," "openly," "manifestly," and "solely."[58] In this light, it might be more accurate to say that "if only the particular be given for which the universal has to be found, the judgment" is *solely*—or can only be— reflective. Reflective judgment possesses characteristics that are uniquely its own and holds a particular and important place in Kant's work. To understand better the distinction between reflective judgment and determinate judgment and to emphasize the imaginative lineage between Kant and American pragmatism, I want to focus on two key aspects of reflective judgment. The first regards the faculty's spontaneity, the second its hypothetical nature.

In the instance of reflective judgment, imagination is described as "self-activating" (*selbttatig*)[59] and spontaneous. Its function in aesthetic judgments is characterized as a type of "lively play."[60] Crawford provides an insightful description of this "play" and highlights the epistemological revision it involves.

> The imagination is in "free play" in the manner in which it gathers together the manifold of intuition (CJ, 9) . . . in the reflective aesthetic judgment, my concern in the gathering operation is not to find a unity which fits some concept or other that my understanding can provide; rather my concern is only whether the organization or

arrangement is such that some concept or other ought to be applicable. In other words, a successful aesthetic reflective judgment is achieved when the experience culminates: "Aha! It—the gathered manifold—exhibits a rule-governed-ness just *as if* it could be subsumed under a concept. It satisfies the conditions for cognition in general."[61]

Crawford's description of aesthetic judgment underlines the difference between these reflective judgments and those described as "determinant" in the *Critique of Pure Reason*. In his earlier work, Kant seems to suggest that the type of investigation implicit in determinate judgments culminates when a unity is formed from the manifold of appearance and subsumed under a pure or empirical rule. In either case the rule is given or preestablished. In the case of reflective judgment, however, no such rule is given a priori. Like the process of gardening, the situation of play supplies the rule and direction for the activity of imagination. The situation suggests what conceptual framework *ought to be possible* in application. As we will see, Peirce and other pragmatists repeatedly emphasize this "Aha!" sensation, the spontaneous and imaginative coalescence of particular observations and possible order, as the basis for and outcome of scientific investigation. This sensation of harmony is crucial to the development of the later Kant. As Förster suggests, "It is in this sensation, this subjective, yet generalizable experience, according to Kant, that demonstrates to us that nature not only harmonizes in its transcendental laws with our understanding but also, in *its empirical laws, it harmonizes necessarily with judgment and its ability to exhibit nature.*"[62]

It is through this imaginative experience that we receive an intimation of natural purposes. The term "intimation" seems to capture the way that reflective judgment points us toward the possible purposes of nature. Kant suggests in the third *Critique* that we can never know, in an apodictic sense, whether an object of our "outer sense" demonstrates natural purposiveness, although we may have very good reason to reflect upon it *as if* it does. For Kant, the inquiry of aesthetic play remains provisional, fallible, and, in a very real sense, hypothetical. The manifold exhibits a structure and dynamic *as if* it could be subsumed under a concept.

At this point, it is important to highlight the quality of imaginative insight. Kant repeatedly states that reflective inquiry reaches culmination, or at least reaches awareness, in an aesthetic *feeling*, a harmony of imagination and understanding. In an aesthetic reflective judgment we *feel* the harmonizing of our cognitive faculties rather than *know* the harmony through concepts. Makkreel and others are hesitant to describe this harmonizing of reflective judgment as just another Kantian "synthesis." As Makkreel notes, "a harmony involves a reciprocal relation between two distinct elements; a synthesis as Kant conceives it, involves a one-sided influence for the sake of strict unity."[63] In the third *Critique*, Kant makes scant use of the term "synthesis" in the discussion of imagination's function in artistic apprehension, primarily employing instead the language of play, harmony, common sense, and feeling. As Makkreel notes, this shift has been overlooked by most commentators, despite the fact that it is a radical departure from the terminology of the first *Critique*, in which all of the functions of the imagination "whether concerning the apprehension of space, the reproduction of images, or the production of schemata—are described in terms of acts of synthesis."[64] Makkreel implies that synthesis suggests a determinant unity while harmony belies a reflective gathering. It still remains to be seen, however, how a judgment of this reflective gathering may be communicated or universalized.[65]

Hooked on a Feeling: Aesthetic Judgment and Common Sense

Some of the obstacles that Kant faces in describing the determinate judgment in the first *Critique* might be overcome in his treatment of the latter in the *Critique of Judgment*. But reflective judgment is not without its own limitations. One of these limitations turns on the issue of the justification and communicability of reflective judgment. Kant emphasizes the role of imaginative mediation in artistic inquiry and the role of a particular common sense in identifying the efficacy of this mediation. At a point in section 20, he even asserts that reflective judgment is a judgment "by feeling." In allowing subjective feeling to provide the awareness of the harmonizing play of the imagination, Kant knows he is treading on rather treacherous philosophic ground. One ought to remember, as Guyer does, that "Kant defines pleasure 'as the idea of the agreement of an object or action with the *subjective* conditions of life.'"[66] One might also recall the difficulty he faces in the first *Critique* when he flirts with the subjective character of knowledge and the inability of this character to be communicated, universalized, or deemed necessary. Kant realizes this danger in the third *Critique*.

> If the pleasure in the given object precedes [the reflective judgment], and it is only its communicability that is to be acknowledged in the judgment of taste about the representation of the object, there would be a contradiction. For such pleasure would be nothing different from the mere pleasantness in the sensation . . . and so could have only private validity.[67]

Again, in the *Critique of Judgment*, by recognizing the role of aesthetic feeling, Kant risks jettisoning any sort of criterion for the apprehension of the beautiful. In section 20, he refuses to succumb to this risk, instead developing a region of taste between "unconditioned necessity" (judgment in accord with definite concepts) and a cognitive realm "devoid of all principles" (evaluation in accord with mere sense). In this middle ground, he insists that the sense associated with judgments of taste is a "common sense."

It is tempting to play fast and loose with the "common sense" of taste. Indeed, Kant himself might be accused of doing so. This being said, it seems wise to address each of the various forms that the *sensus communis* takes in the Kantian corpus. At certain points Kant claims that it is equally a faculty, a feeling, and an exemplary norm. A bit of time should be spent untangling this three-pronged assertion. The three aspects of aesthetic common sense will find a correlation in the various aspects of pragmatic inquiry.

In section 40, Kant introduces common sense not as a sense but as a faculty. It is a common faculty or capacity common to all humans insofar as humans, by definition, employ their cognitive faculties in ordered relation. It is a capacity that harmonizes the cognitive powers in a relation suitable to cognition. More particularly, it is the ability to judge this harmonization in regard to the formal cognitive relations it establishes. In the absence of a determinate concept, it is the ability of the *sensus communis* to evaluate the *formal* relation between the cognitive powers (most notably between imagination and understanding) that allows Kant to maintain that aesthetic judgments of taste are communicable and necessary. Kant describes the *sensus communis* as a faculty in section 40.

Under the *sensus communis* we must include the Idea of a sense common to all, i.e. of a faculty of judgment, which in its reflection takes
account (a priori) of the mode of representation of all other men in
thought; in order as it were to compare its judgment with the collective
reason of humanity and thus to escape the illusion arising from the
private conditions that could be so easily taken for objective, which
would injuriously affect the judgment.[68]

The faculty that Kant identifies is not strictly the ability to evaluate
the play of one's own faculties but a capacity to compare this evaluation
with the possible judgments that ought to obtain in other individuals
under similar circumstances. This hypothetical comparison is done by
"abstracting from the limitations which contingently attach to our own
judgment" and by strictly regarding "the formal peculiarities of our
representation."[69] It is by this comparative process that common sense
becomes truly common. Kant will go on to remark that this ability to
abstract and compare is shared by all human beings but that the power
may differ in degrees in accord with the natural gifts of particular individuals. A heightened common sense can help define the "man of en-
larged thought."[70]

Kant's discussion of the common sense becomes more complicated
when he proposes that this faculty is also a feeling. In section 41, he reminds us that the *sensus communis* substitutes for the definite concepts
described in the first *Critique* that governed the relation between the
understanding and the imagination. As opposed to this determinate situation, when "the Imagination in its freedom awakens the Understanding and is put by it into regular play [harmonious play] without the aid of
concepts, the representation communicates itself not as a [determinate]
thought but as an internal feeling of a purposive state of the mind."[71]
Here Kant suggests that the common sense of taste be associated with a
particular internal feeling, but he is careful to frame this feeling so as to
avoid reducing judgments of taste to the realm of incommunicable sense
or to a mere aesthetic material judgment. He states that this feeling is not
the ground of reflective judgment but rather the consequence of the formal cognitive play that occurs in such a judgment. Early in the "Analytic
of Aesthetic Judgment," Kant writes that the subjective unity of relation
that obtains in reflective judgment "can only make itself known by means

of sensation." The shared feeling of common sense cannot be grounded on experience (as a type of pleasantness or fancy) but on the basis of the form of harmonious mental play. Common sense is described as an embodied sense that is caused by the harmonious play of the faculties. Kant elaborates: "the quickening of both faculties [the imagination and the understanding] which is an indeterminate, but yet . . . harmonious activity, viz. that which belongs to cognition in general, is the sensation whose universal communicability is postulated in the judgment of taste."[72] Common sense is not only the capacity for the cognitive faculties to be in ordered relations but also the bodily sense that makes us aware of this order.

In addition to being a feeling and a faculty, the *sensus communis* stands as a type of ideal or norm. Kant's development of aesthetic judgment presupposes a common sense that serves as a type of benchmark to which all particular judgments of beauty are to conform. It is in this sense that he writes that "in all judgments by which we describe anything as beautiful, we allow no one to be of another opinion."[73] Kant, however, realizes that he may have overstated his case. He reframes his assertion, stating that the norm of the common sense does not dictate that one's judgment *will* conform to its ideal, but that one's judgment *ought* to align with its aesthetic sensibility. The common sense, therefore, should be regarded "as an example of whose judgment I here put forward my judgment of taste and on account of which I attribute the latter [my judgment] an exemplary validity."[74] Kant claims that common sense is an ideal norm but also, and more interestingly, an *indeterminate* norm. If the imagination is to function freely in aesthetic judgments, the standards by which these judgments operate cannot be given in a determinate a priori form under a determinate concept but must remain hooked to common feeling, a common sense.

In the *Critique of Judgment*, Kant exchanges apodictic justification for aesthetic common sense. The *sensus communis* now stands as the *ever-evolving* benchmark for artistic production and apprehension. Instead of retreating from the treacherous philosophic ground of subjectivity, as he does in the first and second *Critiques*, in the third Kant negotiates this region, providing an alternative to a static and conceptual vision of the generalization of judgments: the harmonious sense of aesthetic judgment. This imaginative community shares and shapes history. At least for reflective judgments, Kant comes to realize that his hope for a priori certainty is, very literally, a thing of the past. Like history itself, the artistic community provides a certain rendering of *what is* and subtly directs our attention to *what ought to be*. Drucilla Cornell's description of the *sensus communis* seems particularly appropriate when she writes, "the future nature of this community of the *ought to be* remains open as a possibility in the *sensus communis aestheticus*. It implies a 'publicness' that awaits us, not one that is actually given us, or one that can be given to us once and for all in any predetermined public form."[75] Cornell's interpretation echoes early readings of Kant put forth by Hannah Arendt and other continental thinkers of the early twentieth century.

The *sensus communis* is ever-evolving—meaning that it is simultaneously permanent *and* conditional. It provides both the enabling conditions and limiting factors for aesthetic apprehension and creation. It is in this sense that Kant insists that we "compare our judgment with the possible judgments of others . . . and thus put ourselves in the position of everyone else."[76] Arendt compares this real-world consensus to the movement of transcendental rationality, writing that "impartiality now is obtained by taking the perspectives of others into account; impartiality is not the result of some higher standpoint that would then actually settle the dispute by being altogether over and above the mêlée."[77] This process of comparison is not governed by a determinant rule but is realized intersubjectively by the community of aesthetic taste. It is hypothetical and affective in nature, hypothetical insofar as the result and end of aesthetic judgment, used in both artistic creation and apprehension, cannot be given as a preestablished rule. Kant elaborates on this point. "We could even define taste as the ability to judge something that makes our feeling in a given perception universally communicable without mediation by any [determinate] concept."[78] Here, Kant also insists that the play of the imagination can be *both* subjectively felt *and* universally communicable. That is to say, it can embody realms that Kant earlier—in the *Critique of Pure Reason*—designates as incommensurable. Kant recognizes that the *sensus communis* is in no way a static entity or category of understanding. It is a continuous bridging of the artistic generations that reflects both a kind of historical constraint and spontaneity.

Genius and the Embodiment of the Imagination

This "bridging" between determinacy and spontaneity is perhaps most pronounced in Kant's rendering of the artistic genius. Genius is the ability to produce beautiful works of art. Having developed the supposition of a common sense of taste that could be used in judging beautiful objects, Kant turns to the thorny issue of their novel production as the work of art.

Kant wrote the third *Critique* partly in response to the general mission of the Sturm und Drang, a German artistic movement of the 1770s. This Romantic movement envisioned its mission as freeing culture from what they perceived as arbitrary rules of neoclassicism (derived from French and Latin roots). In contrast, Kant wished to show that some standards are not arbitrary and should not be discarded in aesthetic judgment. This is what he attempts in his discussion of common sense. While Kant wants to characterize the play of the imagination as free, he does not believe that the wild freedom of Romantic genius can support the production of harmonious beauty.[79] Indeed, he often comments that the freedom from all constraints, a merely negative freedom, is in fact a type of self-imposed ignorance. For Kant, true creative freedom requires originality but also a sensitivity to certain circumstances and norms. Freedom occurs in the nexus of both originality *and* communicability. It is this twofold character of freedom that Kant tries to present in his theory of imaginative genius.

Genius's ability to express and communicate aesthetic ideas is particularly interesting in light of the difficulties of reconciling communicability and the spontaneous cognitions of a given subject. By the time Kant writes the *Critique of Judgment*, however, he has hit upon a particular mode of communicable spontaneity embodied in genius that is, properly speaking, spirit (*Geist*).

> For to express the ineffable element in the state of mind implied by
> certain representation and to make it universally communicable—

whether the expression be in speech or in painting or in statuary—
this requires a faculty of seizing the quickly passing play of the
imagination and of unifying it in a concept (which is even on that
account original and discloses a new rule that could not have been
inferred from any preceding principles or examples) that can be
communicated without any constraint of rules.[80]

It is important to note that the "spirit-ed" activities of genius do not
move in accordance with a preestablished teleology. Genius can be "communicated without any constraint of rules." In this sense, the "spirit"
addressed by the third *Critique* should not be confused with the many
hasty interpretations of Hegelian *Geist*. Kant is careful not to speak of a
determinate point toward which this aesthetic discovery and expression
advances, nor of a particular terminus in the evolution of artistic
culture. Such a preestablished teleology would compromise the freedom
that he attributes to artistic genius. In the crucial section 49, the unique
freedom of genius takes center stage.

Genius is the exemplary originality of the natural gifts of a subject in
the free employment of his cognitive faculties. In this way the product of genius . . . is an example, not to be imitated . . . but to be followed, by another genius; in whom it awakens a feeling of his own
originality and whom it stirs so as to exercise his art in freedom from
the constraints of rules and thereby a new rule is gained for art.[81]

This passage deserves detailed attention. First, the obvious points. Genius does not produce beautiful works by rote, memorization, learning,
or example. Kant often notes that genius cannot be taught but rather is
given, inspired. "On this point everyone agrees: that genius must be considered the very opposite of the spirit of imitation."[82] Kant states that the
imagination's free harmony cannot be brought about by referencing the
rules of "science or mechanical imitation" but can only be realized by
the "subject's nature."

As to the not-so-obvious points: We must bridge Kant's appeal to the
originality and novelty of genius. At first glance, at least, the appeal re-
sembles the tendency of the Sturm und Drang movement he opposed so
fervently. A second look at the passage reminds us that genius is the "*exemplary* originality of the *natural* gifts." Since there are no determinate

concepts controlling the process in an act of genius, the rule must come from nature, specifically, nature in the person of the artist. It is in allowing nature, in its organic order, to endow genius with its powers that Kant takes a first step at preserving an originality that will not be confused with original nonsense.

The spontaneous genius stands as a moment of continuity in relation to both the order of aesthetic taste (common sense) and the order of natural beauty. Let us begin by examining this relation in terms of the natural world. It was already mentioned that the play of the imagination as framed in the *Critique of Judgment* was productive and creative rather than merely reproductive. This production is intended and realized by genius. "Genius is the talent [natural gift] which gives the rule to art. Since talent, as the innate productive faculty of the artist, belongs itself to nature, we may express the matter thus: Genius is the innate mental disposition [*ingenium*] through which nature gives the rule to art."[83]

Note the turn of phrase that Kant uses in this passage. Genius *is given* by nature as a natural gift, yet it is precisely through this gift that genius acquires the ability *to give* the "rule to art." *It is through receptivity that genius is productive and spontaneous.* Genius, in this sense, is both passive/receptive and active/self-motivating. Sobel expands on this point, noting that "although 'genius' is productive, Kant is emphatic that this faculty is also *receptive*. It is that through which nature gives the rule to art. Nature acts 'by the *medium* of genius.'"[84] Although it is clear that nature does not operate through genius in a mechanistic fashion, as it operates in the phenomenal laws described in the first *Critique*, Kant suggests that nature acts *as if* it were purposive.

In faithfully translating and expressing the orderability of the natural world in the form of beautiful art, genius is forever bound to nature's structure and emergence. In the thick of things, nature gives genius its cues. Genius responds by reading these cues more or less faithfully. Admittedly, no reading is exact, and no two translations are exactly alike. This being said, however, all readings, to the extent that they are translations, are limited in a certain respect. That the free play of the imagination cannot have a determinate concept as its ruling basis does not mean that the beautiful is free from rules altogether. Nature gives the rule to art though the medium of genius. In this sense, genius is not in, above, or below nature.

Genius is not *apart from* nature. Genius is *a part of* nature. It is only by virtue of its natural endowments that genius can act as genius. It stands as the acting mediator between the ordered beauty of the world and the ordered beauty of art.

Indeed, this faculty demonstrates the continuity of human artistry and the natural world. Kant describes this gifted individual as "author of a product for which he is indebted [*verdankt*] to his genius . . . he does not know himself how he came by his ideas . . . Genius itself cannot describe or indicate scientifically how it brings about its products, and it is rather as nature that it gives the rule."[85] There is always an element of surprise in the inquiry of genius, for an element of the inquiry is always beyond its control. Genius is simultaneously discovering and creating the harmony of the beautiful. This remark gels with the comment that genius is "unsought" and "undesigned" in the sense of craft or artifice.[86]

Despite Kant's emphasis on the originality of genius, the products of genius, inasmuch as they are beautiful, are not wholly free from constraint. First, they are constrained by the gift of nature, but they are also limited by the common sense of taste. Mere genius (spirit) without taste could produce abundant nonsense. Remember that works of genius must be "exemplary." He writes that taste, embodied in the *sensus communis*, "severely clips [genius's] wings and makes it civilized, polished. . . . It introduces clarity and order in the wealth of thought and hence makes the ideas durable, fit for being followed by others and fit for an ever advancing culture."[87] The flight of genius is grounded, at least in part, by the past forms of the *sensus communis*, by taste. Yet this playful faculty still has the force to stretch and challenge these constraints. Indeed, it is the power of genius that expands the notion of aesthetic taste and propels Kant's "ever advancing culture."

The force of genius that expands the notion of aesthetic taste corresponds to Kant's earlier comments on the character of aesthetic ideas. It is necessary to remember the description that Kant gives in the opening of the third *Critique*. He writes that reflective judgments have a reproductive-productive dimension. He adds for the first time that they bear a creative and "spontaneous" element. Kant suggests that in the creation of aesthetic ideas, "we place under a concept a representation of the Imagination belonging to its presentation, but which occasions in itself more thought than can ever be comprehended in a definite concept."[88] He goes on to explain that in this unique type of idea, the imagination "aesthetically enlarges the concept itself in an unbounded fashion . . . and is here creative, and brings the faculty of intellectual ideas into movement." Whereas in the first *Critique* representation always occasioned a definite thought, in the case of judgments of artistic taste, "more thought is occasioned than can be made clear."[89] These comments are unpacked throughout the "Deductions of Pure Aesthetic Judgments."[90] It is precisely these sections to which Rodolphe Gasché turns in order to argue that "Kantian aesthetic judgment is involved in an epistemological, or rather, para-epistemological task."[91] It would seem that artistic genius lies at the heart of this task and will have a central role in the epistemology taken up by American pragmatism. Gadamer makes a similar observation when he employs Kant's notion of the "free play of the imagination" as the cornerstone of his hermeneutics in *Truth and Method*.[92]

The act of genius is executed on the paradoxical cusp of past actualities and future possibilities, in that odd "middle ground" where universality and particularity, determinacy and freedom, hold equal sway. Interestingly, Kant describes this cusp as a type of "happy relation" that the genius enjoys between itself and nature, between itself and the *sensus communis*.

> Hence the genius actually consists in the happy relation—one that no science can teach and that cannot be learned by any diligence— allowing us, first, to *discover* ideas for a given concept, and, second, to hit upon a way of *expressing* these ideas that enable us to communicate to others, as accompanying the concept, the mental attunement that those ideas produce.[93]

Kant's insistence that genius both discovers and expresses aesthetic ideas is important, for it reflects his earlier comments that imaginative genius is both receptive and creative—and that it possesses the talent of aesthetic communication. This passage is unique not only in its rendering of creativity that might support certain pragmatic sentiments but also in the way in which Kant slips between the use of "genius" and the use of "us." It seems quite plausible that genius is a keystone for fine art. But it is also a crucial lynchpin of human discourse and communication.

Genius deserves sustained treatment in the project at hand for the following reasons: (1) Kantian imagination literally comes alive in genius. In "On the Powers of the Mind That Constitute Genius," imagination is embodied, given a human form that is situated in a natural and social context.[94] (2) This embodiment of *ingenium*, genius, stands in relation to the past forms of aesthetic taste as represented in terms of the common sense of taste (*sensus communis*). By virtue of the "free play of the imagination," the genius is able to mediate and translate these past forms, produce novel artistic forms, and thereby extend the legacy of aesthetic sense. The centrality of the imagination in the playing out of what pragmatists will describe as "continuity" gets its first treatment. (3) With his discussion of the beautiful, Kant describes a mode of ordering the manifold of sense without a determinate concept. With his discussion of genius, Kant goes one step further, providing a way of thinking about the novel production of beautiful objects and the communicability of judgments of taste. This way is primarily blocked in the first and second *Critiques*; before the third *Critique*, it remains uncertain whether this sense can be communicated.[95] (4) The character of genius revises the oppositional relation between man and nature. Instead of opposition, the new relation is one of continuity. The development of the aesthetic invites Kant to reframe the ontology he had relied upon in the *Critique of Pure Reason*. As Gasché and Förster note, genius, a cognitive disposition that human individuals have in varying degrees, is a natural gift.[96] Genius is *of* nature. Kant expands on this point, writing, "the innate productive faculty of the artist belongs itself to nature. . . . Genius is the innate mental disposition through which nature gives the rule to art."[97] The freedom of imaginative genius is tied to or, more appropriately, *rooted in* a natural setting. In this sense, the genius mediates between subjective feeling and formal order but also and at the same time between the natural and the ostensibly human.

As I will show later, these four aspects of genius are taken up by C. S. Peirce not in a limited aesthetics but rather in a comprehensive epistemology and ontology. Peirce will take his cues from Kant but will go further than Kant was willing or able to go in fostering the growth of an imaginative philosophy.[98] If successful, the discussion of Kantian imagination will prove roughly isomorphic with the analysis of Peirce's epistemology/

ontology in the second section of this project (chapters 3 through 6) and with the investigation of the empirical sciences in the third (chapters 7 through 9). The movement and development of the imagination in Kant will anticipate and, in a certain sense, map pragmatic and empirical studies of cognition.

Opus Postumum: *From the Imagination to a Natural Ontology*

Before turning to Peirce, however, a last comment is warranted with regard to Kant. After his description of reflective judgment in the third *Critique*, Kant dedicates himself to a revision of the natural sciences. This turn is unsurprising when one considers the way in which aesthetic genius (which figures centrally in the development of the *Critique of Judgment*) acknowledges and expresses the purposiveness of the natural world. In 1792, Kant writes that "[reflective] judgment first makes possible, indeed necessary, for us to think of nature as having not only a mechanical necessity but also a purposiveness; if we did not presuppose this purposiveness, there could not be systematic unity in the thoroughgoing classification of particular forms in terms of empirical laws."[99]

While far more attention is warranted in regard to this comment, it is clear that Kant's philosophical stance foreshadows a key philosophical presumption of classical American thought—that nature must be of the same order, must have the same structure and form, as purposive thought. Peirce restates this presumption when he insists that science depends not only on the isomorphism but on the continuity between purposive thought and the purposes of nature. For Kant, however, this principle, as a principle of reflective judgment, remains only speculative in nature. Peirce appropriates this insight into the character of natural purposes and makes it the lynchpin of his epistemology and ontology. In examining the conditions for the possibility of pragmatic inquiry, Peirce will echo the later Kant in his belief that such inquiry, especially the processes of abduction, presupposes the purposiveness of nature.

In the preface to the *Critique of Judgment*, Kant states that "with this, then, I bring my entire critical undertaking to a close."[100] Kant's later work in the *Opus Postumum*, however, indicates that this statement was somewhat premature. The ostensible end of Kant's critical project draws

him into a full-fledged investigation of the natural world as being continuous with the imagination, the ordered and synthesizing relation that
he identified two decades earlier in his development of the first *Critique*.
While a detailed exegesis of Kant's last work is not possible in the context
of the current project, it would be negligent to overlook Kant's development of the connection between imaginative judgment and the realization of nature's purposiveness. Kant begins to outline this connection
in the *Critique of Judgment*. Through reflective judgments, "independent
natural beauty reveals to us a technic of nature that allows us to represent nature as a system in terms of laws whose principle we do not find
anywhere in our understanding."[101] The imagination, operative in reflective judgment and, more particularly, in teleological reflective judgment,
has the task of securing a type of intelligibility for "phenomena of nature
that the understanding not only cannot explain but cannot even bring
into view as such."[102] This intelligibility comes home to us when we judge
the manifolds or organized forms of nature guided by the principle of a
natural purpose. Once again, it is important to note that we do not *understand* this purpose as a determinate concept. Rather we are invited *by
nature* to judge and act as if such a purpose were at work in the phenomena. The imagination, in effect, "expands" our concept of nature beyond
a mere understanding of natural phenomena and grants us the ability to
be in touch with the possible ends of nature.

It is widely accepted that Kant's understanding of nature begins to
shift in his later works, a fact that seems to correspond with his studies of
the aesthetic element of cognition. Instead of the presenting the blind
and mechanistic rendering of nature seen in the first *Critique* and the
Metaphysical Foundations, Kant begins, in the words of Förster, to envision "nature as art, hence to a nature that is in itself systematic."[103] Much
of Kant's work in the *Opus Postumum* aims to highlight the relation
between this natural system and human consciousness, developing the
point that he makes at the end of the *Critique of Judgment*. He repeatedly
argues that a human being is continuous with the natural world, and, in
its organic-bodily movement and interaction, the human being reflects
the organization and purpose of nature. There is much debate concerning the direction of influence between the *Critique of Judgment* and the
Opus Postumum. But it is safe to say that these remarks resonate with

earlier comments made on the nature of genius in the third *Critique*: genius is a natural gift. Genius naturally embodies a relation that is both receptive *and* active. By 1796, Kant begins to reconsider the bivalent disposition of genius, or, more accurately, he begins to attribute this disposition to human inquiry on the whole. He suggests that as imaginative and embodied thinkers we are "spectators and, at the same time, originators."[104] This project extends this twofold character of imaginative inquiry and gives it a central place in development of American thought in the following century.

THREE

C. S. PEIRCE AND THE GROWTH
OF THE IMAGINATION

The Seeds of the Aesthetic

The seeds of the aesthetic are buried deep in the work of Charles Sanders Peirce. The sprouts were, therefore, rather slow to show themselves. However, the concept of the imagination as framed by Kant and other German Enlightenment thinkers does emerge in the ground of his epistemology, ontology, and metaphysics. Peirce recognizes the important function of Kantian reproductive imagination as he develops a description of inquiry that underscores the continuity between bodily sensation and understanding, once again challenging the mind-body dualism to which the idealists and empiricists continually fell prey. A more radical claim emerges from this challenge, namely that the "genius" of pragmatic inquiry, like the genius of Kantian creativity, ought to be regarded as a "gift of nature," as being continuous with and arising from the natural processes of organic life. The continuity that is highlighted between the sensuous and the conceptual implies a certain unity between the material reality and the mental lives of individuals which will be highlighted in coming chapters. In addition to addressing the mediating character of the imagination, Peirce seizes upon the character of the productive and creative imagination, translating the aesthetic concept into the language of epistemology and ontology. Peircean abduction, the process of hypothesis generation, dramatically demonstrates this translation: Peirce takes up the issues of spontaneity, growth, and adaptation as he advances his logic of abduction.

The two defining characteristics of Peirce's philosophy—spontaneity and continuity—are the primary subjects of the next three chapters. In concentrating on these two aspects of Peirce's thought, it is possible to argue that the logic and ontology of pragmaticism are defined by the aesthetic imagination. More specifically, I will argue that abduction, a hallmark of Pierce's philosophy, is a direct offshoot of this aesthetic process as it has been described vis-à-vis the thinking of the German Enlightenment. Peirce extends the concept of the imagination outlined by Kant, more or less unknowingly, in order to place philosophy on a new footing.

If the imagination is at the core of Peirce's thought, it is often covered over or overlooked in contemporary Peirce scholarship. My investigation takes its cue from Phillip Weiner's claim that Peirce "boldly generalized the role of the imagination" and described it as the "source of all the sciences."[1] Weiner's framing of the imagination, however, did not allow him to examine the implications of his initial observation. My account is indebted to Douglas Anderson's work in *Creativity and the Work of C. S. Peirce*, where he presents a sustained argument that there is a latent theory of aesthetic creativity in Peirce's extremely analytic corpus.[2] At points, it echoes a portion of Murray Code's analysis of Peirce's appropriation of the Kantian project, for example when he writes that "Kant's inability to tame the imagination, and thus live up to his promise to reform metaphysics once and for all, invites a return to his original insight—that there is an originary imagination involved in the very constitution of experience." Peirce provides a way of thinking through the creative imagination as the constitutive element of consciousness on the whole.[3]

Framed in this way, this section of the project also emerges as a rebuttal to scholars who continue to downplay the importance of aesthetics in Peirce's work. Beverly Kent, for example, treats aesthetics as a small subsection of Peirce's categories and claims that, "Peirce developed an increasingly clear idea of the sort of science aesthetics must be if it was to harmonize with his architectonic development."[4] Instead of suggesting that Peirce had to find a place for aesthetics in his philosophy, I will argue that the concept of aesthetic imagination constitutes the very foundation of his thinking. I would agree with Kent that Peirce's explicit comments on the nature of aesthetics are "permeated with anomalies and sometimes with bizarre assertions."[5] Despite this, his rendering of human cognition, ontology, and metaphysics is uniquely artistic in character and helps us recognize the motivating force that a theory of imaginative thinking can have in the doing of philosophy.

This reading of Peirce also stands against accounts that attempt to enlist particular Peircean notions into the service of aesthetic theory. Albert Levi, for example, provides a cursory argument for the employment of the Peircean categories of firstness, secondness, and thirdness in formal art criticism. He believes, perhaps quite rightly, that these categories can respectively account for the qualities, the movement, and the meaning of a work of art.[6] In the 1980s, Carl Hausman began to advance a philosophically sophisticated argument that Peirce's semiotic could help explain how a work of art "is fitting or appropriate not only to itself, but to the world 'outside,' that is, to a world outside the work."[7] Peirce's notion of an indexical sign allows Hausman to suggest that there is a real and affective reciprocity between a work of art and the object to which it refers. While their respective treatments of Peirce's work seem well conceived, scholars such as Levi and Hausman, with their unique selections from Peirce's corpus, tend to downplay the way in which his *entire* body of work might be read through the lens of the aesthetic. I seek to address this neglect. At the very least, if successful, these chapters will serve as a radical extension of these more limited arguments on the aesthetic impact of Peirce's thought. It should be noted that Richard Smyth's reading of Peirce, in *Reading Peirce Reading*, comes close to accomplishing this task. His analysis, however, often focuses on the correspondences between Kant's second *Critique* and Peirce's philosophy, developing what Smyth calls Peirce's "second Critique strategy."[8] Smyth repeatedly asserts that Peirce did, in fact, pay great attention to the *Critique of Judgment*, but I have not been able to find documentary evidence to this effect. I hope to show that this fact does not jeopardize our tracing out of the imagination in Peirce's writings. If further archival research reveals that Peirce did in fact read the third *Critique* first, as Smyth suggested, it will not, I believe, undercut my project in any significant way.

In turning our attention to Peirce, let us state the obvious: his expansive corpus is notoriously difficult to negotiate. In the attempt to provide a unified theory of thinking and being, Peirce negotiates wide-ranging philosophical fields: logic, epistemology, ontology, and metaphysics. For this reason, topical approaches to his work often fall short in their explanatory power. In avoiding this pitfall, I will adopt a chronological reading, occasionally employing *Writings of Charles Sanders Peirce: A Chronological Edition.* Unfortunately, this series only consists of Peirce's early works prior to 1890. In certain cases when this edition could not provide adequate materials, I have turned to the *Collected Papers of Charles Sanders Peirce*, an edited, topical collection that is more comprehensive but more difficult to manage. In many cases, I have employed the unpublished manuscripts housed at the Houghton Library at Harvard University. This type of historical investigation exposes the maturation of Peirce's thought and the way in which the ordering structure of the imagination takes on a larger role in Peirce's later thinking. A similar movement has been highlighted in the development of the prior chapter on Kant and the imagination.

Peirce's Early Thought: The Aesthetic Letters *and Genius*

A cursory examination of Peirce's early work (1853–1871) exposes two basic points: (1) At an extremely young age, Peirce acknowledged the importance of the imagination as developed by Friedrich Schiller and, by proxy, Immanuel Kant. (2) Peirce's acknowledgment, however, did not grant him the insight to avoid fully the problems that Kant himself had faced in the *Critique of Pure Reason.* This fact is reflected in his early forays into triadic logic. It is fairly widely acknowledged that Peirce's notion of continuity, a notion that gels with the imaginative account that my book advances, did not come into view until the late 1880s.[9] His early descriptions of true continua fall short in their unwillingness to address the unique mediation between matter and mind and the character of genuine spontaneity.

At the age of fourteen, Peirce wrote that "Poets see a common nature."[10] This aesthetic sensibility was deepened two years later, in 1855, when Peirce began to read Schiller's *Aesthetic Letters*. Schiller's analysis of the beautiful and genius, an analysis widely accepted as an offshoot of the descriptions provided in the third *Critique*, suggests that there is a discrete process that mediates between sensibility and understanding. Schiller occasionally calls this middle mood "the beautiful"; at other points, he calls it the imagination. In his first published essay, "The Sense of the Beautiful," Peirce writes that Schiller has rightly identified a "third impulse which creates beauty" between the "sensuous impulse" and the "rational impulse that gives laws."[11] Here, Peirce inadvertently has stumbled across Kantian reproductive imagination, described in some detail in the previous chapter. In his early writings, including his 1859 "Analysis of Genius," he develops a portrait of genius wholly congruent with the rendering given in the *Critique of Judgment*.

Like Kant and Schiller, Peirce maintains that genius is "a mind of large general powers and admits that his mind may receive accidental determination." Peirce also "asserts that the determination is likely to be in great degree original."[12] The work of genius is equally attributable to accident and autonomous activity. He echoes Goethe in writing that "there is a sense in which it is true that the poets and real artists must be born, not made. Namely there must be an inward productive power, to bring the images which linger in the organs, the imagination, the memory, freely without purpose or will to life."[13] That artistic genius "lingers in the organs" and must be drawn out in this purposeless purpose is a conception of a process that resonates very clearly with the description of the imagination in the previous chapter. In describing the relation between the natural and the artistic, Goethe is responding to and extending Kant's aesthetic theory. Peirce, in turn, is involved in a similar project. The young Peirce explains that men and women of genius "must unfold themselves, grow, extend, and accumulate in order to become instead of figurative diagrams really present pictures."[14] This organic process of becoming is not a matter of fabrication or imitation but rather, as Kant suggests in the *Critique of*

Judgment, a matter of natural unfolding in which the genius participates but which it does not fully control. The genius's participation is not solely determined by his or her willful decision but, as a gift of nature, is partially given or compelled. Peirce writes:

> Another trait observable in famous poets is an instinctive invisible impulse to express the ideas and feelings within them; the consequence is that the work of genius is not a voluntarily labored product, but the involuntary product of psychical need. It is not a hankering after applause and success nor a regard for his interests which make the artist of genius work. It is solely a hankering to give shape to the work of art that exists in his mind. The true poet does not testify because he will, but because he must. Goethe has painted the poet impulse in Tasso.[15]

Here, Peirce draws heavily from the German aesthetic tradition but also more directly from Emerson's suggestion that the genius must be attuned to the minute nuance of experience and able to respond to this attunement. In an early undated manuscript, Peirce elaborates on the concept of genius, agreeing with Alexander Gerard's 1759 account in "An Essay on Genius." Without attributing the quotation, Peirce transcribes Gerard's position that genius is

> an extensive comprehensiveness of the imagination, in a readiness of associating the semiotist [*sic*] ideas that are in any way related. In a man of genius the uniting principles are so vigorous and quick, that whenever an idea has been present to the mind, they bring into view at once all others that have the least connection with it.[16]

This passage is remarkably similar to the stance that Kant assumes in the *Critique of Judgment.* It is by way of the imagination that human cognition achieves coherence, but it is also by way of the imagination that new conceptual forms arise. Here, Peirce seems to suggest that the imagination is never simply reproductive (as described in the *Critique of Pure Reason*) but rather is bound up in the associative process of re-creation, generation, or "invention." Peirce's early interest in genius and the imagination would linger and inform his later writing. In 1896, for example, Peirce translates and reviews William Hirsh's *Genius and Insanity* and redeploys Gerard's position on genius and the imagination in his critique of the author's argument. Peirce refers again to the Gerard quotation above and adds this: "A more pregnant utterance would be hard to find. The theory of association here implied and its application to genius might afford a man's reflections enough to occupy a long sea voyage."[17]

The similarity between the aesthetic theories of Gerard and Kant is not coincidental. Indeed, the similarity follows a strand of eighteenth-century intellectual history that Piece would extend in the development of American philosophy in the coming century. Gerard dramatically influences Kant's later writings. The affinities between Kant and Peirce, therefore, should not only be traced to Peirce's reading of Friedrich Schiller (an intermediate link between the two thinkers) but also to Gerard, a Scottish thinker whose writings on genius and the imagination inspired both Kant and Peirce. Kant admired the rendering of artistic genius developed by Gerard, stating that he was the best English-speaking writer on the subject: "*ein Englander, hat von Genie geschrieben, und darüber die besten Betrachtungen angestellt obgleich die Sachesonstauch bei anderen Schriftsstellern vorkommt.*"[18]

In light of the connection among the three thinkers, it is truly a shame that Peirce was not directly exposed to Kant's aesthetic theory (or, if he was, made more of this exposure). Despite the apparent similarity between Peirce's early observations and Kant's later work, it seems very unlikely that Peirce spent a great deal of time with the third *Critique*. He did work extensively with the *Critique of Pure Reason* and in 1858 undertook a translation of the first edition with the help of his aunt.[19] His letters to Francis Abbot in March 1865 indicate a "personal enthusiasm for Kant."[20] But he fails to mention any knowledge of Kant's later works that round out the descriptions of the imagination or genius. Over the course of one month, this deep and inspiring "enthusiasm" seems to have turned to a deep dissatisfaction with Kant's project in the first *Critique*. By the time of his "Lecture on Kant" later that year, Peirce states: "Every man who wishes to vindicate his pretensions to philosophic power must display it by the discovery of an error in Kant."[21]

With Abbot's help, Peirce does discover an error, namely that Kant reinstantiates the form-matter divide of modern philosophy in the idealistic separation of rationality and sensibility. Furthermore, Kant ignores the problem implicit in the notion of things-in-themselves.[22] Abbot would summarize their criticism of Kant in *The Syllogistic Philosophy* (1903), writing that they intended to "sweep away that doctrine of the necessary separation of sense and intellect, of experience and reason, which lies at the very foundation of idealistic philosophy . . . notably as illustrated in the 'pure reason' and 'pure *a priori* knowledge' of Kant."[23] On this note, Peirce takes Kant to task for not recognizing the vital importance of the schemata, the one mediating process that might overcome the epistemo- logical separation of sense/experience and intellect/reason. Peirce writes, "the doctrine of the schemata can only have been an afterthought [for Kant], an addition to his system after it was substantially complete. For if the schemata had been considered early enough, they would have over- grown his whole work."[24] Peirce's reading of the schemata as the ne- glected keystone of Kant's epistemology and ontology is extremely similar to the reading advanced in the first two chapters of this book. Interest- ingly, it should now be clear that Peirce appropriates Kant in content and language when he writes in the late 1860s that "it remains true that there is, after all, nothing but imagination that can ever supply [one] an in- kling of the truth. He can stare stupidly at phenomena; but in the ab- sence of imagination they will not connect themselves together in any rational way."[25] To say that Peirce appropriated Kant is not to say that Peirce is Kant warmed over. Indeed, by 1911 in his unpublished "A Logi- cal Interpretation of the Christian Creed," Peirce asserts that his early interpretations of the *Critique of Pure Reason* are far clearer, far more original, and far sounder than Kant's own thought.[26]

Young Peirce Righting Kant's Wrongs

If we view his work in light of his critique of Kant, we must assume that Peirce's pragmatic philosophy *was* "overgrown" or, at least, defined by a schema-like process of mediation that is enacted by the imagination. This assumption is borne out by three essays written in 1868: "On a New List of Categories," "Questions Concerning Faculties Claimed of Man," and "Some Consequences of the Four Incapacities." Hausman and others contend that Peirce's pragmaticism was born in 1877–1878 with the writing of "The Fixation of Belief" and "How to Make Our Ideas Clear," respectively.[27] Yet many of pragmaticism's characteristics are on display in these 1868 essays.

In the first of these pieces, "On a New List of Categories," Peirce revises the Kantian categories in an attempt to explain how human conception "reduces the manifold of sensuous impressions to unity."[28] He concentrates on the process of mediation that must occur so that pure concepts are able to have empirical content. His treatment antici- pates his post-1880 insistence that the categories be developed phenom- enologically. In "On a New List of Categories," Peirce takes his first look at the philosophic significance of the copula's conjunctive relation. The copula functions as a medium, a connective tissue, between two propo- sitions and allows for the possibility of continuous thought. (In later work, Peirce highlights the copula's imaginative character and repeat- edly reminds us that the copula itself is the "conception of being."[29]) Here Peirce also takes a first swing at a theory of signs, underlining the importance of "the third" as "an interpretant."

In "Questions Concerning Faculties Claimed for Man," the triadic character of the copula emerges as a primary topic. In this case, however, Peirce elaborates on the "proposition" of "the third" in the broader dis- cussion of continuity, in an attempt "to prove and trace the consequences of certain propositions in epistemology tending toward the recognition of the reality of continuity and generality and showing the absurdity of indi- vidualism and egoism."[30] This attempt is partially successful and fore- shadows his development of "synechism."[31] Peirce's main point is that we, as thinking individuals, *are in touch* with the material world. This sugges- tion stands in stark contrast to the mind-body dualism that continued to define the philosophic plot in which Peirce would sow his pragmaticism's seeds. In this essay, Peirce moves in typical form from a discussion of logical propositions to a description of human inquiry that is embodied and rooted in sensuous experience. He presents a sustained explanation of, and direct argument against, Kant's work in the Transcendental Aes- thetic (the first *Critique*) where Kant describes the problematic relation between bodily sense and understanding.[32] Peirce's rendering is a prepa- ratory move to his proposition of the organic basis of cognition, and it begins to mesh with his earlier descriptions of Kantian genius as being a *natural* gift. This sets the stage for Peirce's naturalized epistemology.

At this juncture, I point to the commonality of this essay with a section of an 1879 essay entitled "Thinking as Cerebration." The 1879 piece elucidates certain suggestions in the earlier work, restating the imaginative nature of thinking and, by extension, the physical substratum of thought. Here, he writes that "thinking is done with the brain and the brain is a complexus of nerves; so thinking is necessarily subject to the general laws of nervous action."[33] The 1879 "Thinking as Cerebration" stands as an extremely early attempt to discuss the organic embodiment of thought. Peirce, having criticized Kant for drawing too hard a line between observation and ratiocination, refuses to make the same mistake in describing the bodily and sensible basis of human thought. This point is made emphatically when Peirce asserts that "we meet no sure indications of a consciousness unconnected with a nervous organism and the more complicated the organism, the higher the consciousness."[34]

"Some Consequences of the Four Incapacities" rounds out this 1868 selection. Peirce later claimed that this essay was one of his strongest pieces of writing. Perhaps he correlated "strength" with extreme density of prose. At points, this work is nearly impenetrable. Buried in this short essay, however, we find germinating seeds of Peirce's mature thought. Four general points emerge in the piece that deserve recognition, points that take their cues from the two preceding 1868 essays and direct us to a way in which we may categorize his later thought.

First, Peirce highlights the ordered mediation that must take place in the continuity of bodily sensation. This mediation is addressed more fully in coming sections. But several of his comments stand to be unpacked as a preliminary matter. Peirce suggests that thought is inextricably bound to phenomenal experience and that it develops continuously as a train (of thought) over time. He writes in a style that William James would come to describe as stream of consciousness:

> If after any thought, the current of ideas flows on freely, it follows the law of mental association. In that case, each former thought suggests something to the thought which follows it . . . there is no intuition or cognition not determined by previous cognitions, it follows that the *striking in of a new experience* is never an instantaneous affair, but is an event occupying time, and coming to pass by a continuous process.[35]

This description of the development of thought recalls the earlier comment that the imagination is that process by which the old is made anew. With it, Peirce anticipates John Dewey's *Art as Experience* by noting that the "striking in of a new experience," a dynamic that characterizes imaginative thought, is always situated in the continuous processes of history and nature. Imagination grants the possibility of both continuity and novelty. In light of our earlier descriptions, it seems reasonable to suggest that the imagination underpins mental association, thereby setting the stage for the recreation and reinterpretation of mental images (Peirce occasionally refers to these as "thought-signs"). Only by virtue of the imagination "is every thought-sign ... translated or interpreted by a subsequent one, unless that all thought comes to an abrupt and final end in death."[36] As Gasché underlines in the work of Kant, one can understand the imagination, the principle of ongoing association, as the animating force in cognition.[37] This point seems to hold in reference to the pragmatists' early investigations of epistemology.

As a second point, Peirce's description of thought as an embodied event that lives and takes place continuously over time leads him to draw out a second important point in the "Four Incapacities." For the first time, he describes, in a surprisingly phenomenological manner, the continuous relation between particular sensations and conceptions over time. The terminological slippage between "sensation" and "conception" that occurs in this essay is significant since Peirce wants to emphasize not only the continuity of thought over time, a continuity that he and Kant both agree is secured by the imagination, but also the vital connection between human sensation and complex modes of conscious inquiry. For Peirce, every thought is a feeling-thought, a conception shot through by a corresponding emotion that obtains in the body of an organism. It is in this respect that he writes that "every thought, however artificial and complex, is, so far as it is immediately present ... a sensation without parts" and states that thought is always accompanied by an immediate unanalyzed *quality* of feeling.[38] The ebb and flow of feeling grounds abstract conception in the sense that it corresponds and co-emerges with the bodily and biological rhythms of an organism. This feeling rhythm is not merely additive to thought but constitutive and even determinative in cognitive judgments. Note that this point is brought out by Peirce in a brief comment on the sensation of the beautiful. Peirce unknowingly echoes Kant's sentiment that aesthetic judgments are made imaginatively, by way of a process that is both affective and, in a certain sense, intellectual.

The third and fourth points made by Peirce in the "Four Incapacities" are made only in passing, but they prove vitally important to an understanding of his later thought. In this early work, human inquiry begins to be portrayed as having access to "the real," which Peirce here equates with "the true." This portrait stands in marked contrast to those accounts that set human cognition against the reality it seeks to investigate. This point resonates with my earlier suggestion that Peirce "abjures from the bottom of his heart" the Kantian thing-in-itself as the unknowable ground of human cognition. Like the works of genius, intellectual investigations are not set against the backdrop of an alien world but rather move continuously with the world at large. Inquiry happens as a continuous event in the intimate coupling of the individual and its environs. As the process of abduction will demonstrate, it is only through this coupling that such investigations can disclose the workings of reality.

Finally, Peirce elaborates on the nature of this reality, underlining the way in which the real is exposed and disclosed over time by a community of inquirers. His elaboration makes explicit the extent to which knowledge is communicable and accumulative. He writes:

> The real, then, is that which, sooner or later, information and reason would finally result in, and which is therefore independent of the vagaries of me and you. Thus the very origin of the conception of reality shows that this conception essentially involves a COMMUNITY, without definite limits, and capable of an indefinite increase of knowledge.[39]

Each of these four points resonates with remarks made earlier in reference to the character of aesthetic judgment, the disposition of genius, and the development of the aesthetic *sensus communis*. This echo of the imagination will be drawn out in detail in Peirce's epistemology. The four general points of "Four Incapacities" also set the stage for a more detailed discussion of Peirce's investigation of human inquiry in the 1870s and 1880s.

Peirce's Refining Thought: Continuity, Time, and Feeling

During the 1870s, Peirce seized upon the importance of continuity not only as a phenomenological characteristic of human thought but as a logical and ontological concept. His reading of Duns Scotus encouraged him to commit to the conception of generality as proposed by the scholastic realists. As Hausman notes, this commitment should not be confused with an acceptance of the reality of universals, as they are usually conceived as being static and determinate. Instead a "general" is a dynamic tendency for growth that cannot be understood apart from an end or *telos.* "Generals can grow," Hausman explains, "first individually, by changing identity or by being modified, and, second, as complexes of intelligible identities or rules that contribute their intelligibilities to an evolving system of generals."[40] Peirce distinguishes his position from his understanding of Hegel's universal to the extent that the ends of this evolving system will never exhaust the novelty (firstness) and friction (secondness) that continually show themselves in the dynamics of the universe. Peirce states this clearly in his notebooks:

> The particular proposition of scholastic realism which most vitally concerns pragmaticism is, however, not an answer to the question whether there are any real generals, which is the same question as is there any real necessity, but it is the proposition not explicitly covered by the usual definition of scholastic realism, namely that there is real *possibility.*"[41]

This concept of generality and continuity as intimately wedded to possibility underpins Peirce's metaphysics. More immediately, the concept serves as the touchstone of his work in logic and epistemology.

Along these lines, Peirce extends his understanding of the relation between time and inquiry, suggesting that the continuity of time must be taken into account in the development of formal logic. This suggestion is poignantly expressed in Peirce's "Time and Thought" (1873) and "Time as Fundamental Character of Formal Logic" (1873). Peirce's insistence that formal logic must describe human experience, an experience situated in the temporal world, forces him to revise the typically atemporal character of dyadic logic. The "logic of relatives" that Peirce begins to explore in the early 1870s relies on his investigations of the nature of time. In these essays, he begins to provide a detailed account of logical "thirdness," a notion that coheres with the descriptions of imagination presented by this book.

It should be emphasized that this discussion of time is in no way tangential to our broader goal of thinking through the imagination. The relation between the intuition of space and time and the imagination's reproductive and productive roles should be clearly in view at this point of our analysis. It is helpful to remember Kant's ambiguous assertion in the section on the Transcendental Analytic that the imagination, that odd *natural* gift, creates the possibility of temporal order and progression. This distinctive capacity of the imagination is also developed in Kant's rendering of the unity of apperception. Peirce may well clarify the general assertion in the Transcendental Analytic. At the very least, he offers us a pragmatic frame in which to analyze Kant's claims in regard to the relations between imagination, time, and human thought.

By placing time at the heart of his logic, Peirce not only introduces an idea of continuity into his epistemology but the concepts of futurity and spontaneity as well. These concepts deserve attention, for only through them can one understand the full import of Peirce's subsequent development of abduction, the process by which new elements arise in human inquiry. In addressing spontaneity of thought, Peirce seems once again to appropriate unwittingly the structure and function of the imagination in order to place logic and philosophy on a new course.

In the summer of 1873, Peirce writes that "the significance of thought lies in its reference to the future . . . thought is rational only so far as it recommends itself to possible future thought."[42] This comment, written in the midst of Peirce's extensive study of logic, belies his suspicion that logic must describe and be an instantiation of a particular type of human inquiry. It is an inquiry that is historically continuous, creative, and spontaneous. Finally and significantly, it is an inquiry that stands in relation to the possibility of future events. Peirce tries, with varying degrees of success, to seize upon this form of inquiry in the "Doctrine of Chances" (1878) and the "Probability of Induction" (1878).[43] In the hope of constructing a logic that is faithful to human, temporally oriented inquiry, he concentrates on a theory of probability that opens the way to a novel form of quantified logic. As Peirce himself states, "the theory of probability is simply the science of logic quantitatively treated."[44]

For Peirce, quantification was one of the first attempts to solve a problem that had long plagued symbolic logic: the problem of negotiating the particular patterns and complexity of observable events by way of simple logical operators. Aristotle recognizes this difficult problem in *De Anima*: "If there is no single common method for the investigation of particulars, then putting our inquiry into practice becomes more difficult. For we will have to grasp *in each case* what the method of inquiry should be."[45] Aristotle was right. Putting our inquiry into practice becomes much more difficult in the absence of a determining method. Like many logicians before him, Peirce realized that universal and even particular logical statements, traditionally conceived, lacked accuracy and, perhaps more importantly, predictive power. They could not provide a common method of inquiry and common modes of inference for unique situations over time. Universal statements reflect the "two conceivable certainties with reference to any hypothesis, the certainty of its truth and the certainty of its falsity."[46] These two certainties serve as the extremes in Peirce's theory of quantification and are designated 1 and 0 respectively. These certainties, however, rarely obtain in the process of human inquiry—they only serve the function of limits. Instead, inquiry proceeds by way of relative frequency in order to evaluate given hypotheses and in order to make particular inferences. In light of this fact, Peirce exposes a need to refine particular logical statements (I and O statements) in order to increase their explanatory force. By quantifying these statements, that is, by assigning each statement a particular numeric variable between 0 and 1 that approximated the frequency of occurrence, Peirce was able to describe more closely the relations (conjunctive, disjunctive, and implicative) of the empirical world as they develop over an extended, albeit definite period of time.

The investigation of logical quantification and the inductive processes that it attempts to model, however, did not satisfy Peirce's desire to describe the novelty and growth of human cognition. Probability and quantified logic turn on the issues of prediction and inference but cannot anticipate the genuine newness that, by definition, could not be described in terms of frequencies generated from the observation of events. Something was still missing in his logical system.

Peirce finds what he is looking for in hypothesis formation, in a process that he will later term "abduction." As many scholars have noted, Peirce confuses induction and hypothesis in his pre-1870 work.[47] One, however, should not be overly critical. The young Peirce is in good company in his mistake: Bacon, Whewell, and Mill all flirted with the conflation of induction and hypothesis formation.[48] In "Deduction, Induction, and Hypothesis" (1879), however, Peirce divides the modes of synthetic inference into the categories of induction and hypothesis.[49] These two stand against the analytic judgment of deduction. In the case of induction, one generalizes from a number of cases over time of which something is true and infers that the same thing is true of the whole class. Kant and Peirce concur that this process is synthetic. Indeed, these remarks on Peircean induction share more than a family resemblance with Kant's discussion of the unity of the manifold of sense. Induction gathers from a series of observable facts and, by habit, generalizes from this gathering.

Hypothesis stands opposed to the habitual process of induction in that it proceeds not from observable fact but from the speculative arena of analogy, similarity, and difference. In other words, "induction infers from one set of facts, another set of similar facts, whereas hypothesis infers from the facts of one kind to another."[50] I will later argue that insofar as hypothesis is spontaneous in structure, it is imaginative. Metaphors, for instance, provide a type of meaningful mediation between separate classes of predicates, allowing humans to infer heuristically from a variety of experiences. As Douglas Anderson asserts, "creative metaphors may have an agential source in the Peircean system."[51] This suggestion will receive a detailed treatment in the final chapters of this book and more immediately in the coming section on the logic of abduction.

It is important to note that the cognitive mediation involved in hypothesis formation is co-emergent with a particular knowing-feeling. Peirce observes that "there is a peculiar sensation belonging to the act of thinking that each of these predicates inheres in the subject." The feeling of a hypothesis, however, is not merely the sensation that is associated with the thinking of a particular predicate. He explains it in this way:

In hypothetical inference this complicated feeling so produced is replaced by a single feeling of greater intensity, that belongs to the thinking of the hypothetical conclusion. Now, when our nervous system is excited in a complicated way, there being a relation between the elements of the excitation, the result is a single harmonious disturbance which I call an emotion . . . this emotion is essentially the same thing as an hypothetical inference and every hypothetic inference involves the formation of such an emotion.[52]

To examine the imaginative legacy of this comment, we need only to return, once again, to our discussion of Kant and to Peirce's explicit remarks on aesthetic judgment and the imagination. The imagination is an embodied and feeling process by which the manifold of sense is brought into harmony without an appeal to an ordering determinant concept. The Kantian schema is co-emergent with a particular feeling of suitedness and harmony; this fact is drawn out in the discussion of moral judgment in the *Critique of Practical Reason* and given the limelight in the third *Critique* in the sections dealing with the feeling relation of aesthetic judgment.

With hypothesis, Peirce hits upon this feeling relation of cognition. As Thomas Alexander notes, Peirce describes the suitedness of a good hypothesis as the harmonious relation that mediates between a community of inquiry and the world at large.[53] Furthermore, Peirce discovers a process that mediates between the impressions of sensible predicates and the *possible* formation of concepts. It is important to note that hypothesis is imaginative not only in its function of mediation but also in its spontaneous production. Hypothesis is "a perilous and bold cognitive step," yet it is a necessary one in the creation of novel thoughts. Peirce observes that induction reflects the habitual aspect of cognition but that it fails to account for the creativity of thinking. Deduction merely works out the formal constraints of a closed system and contributes nothing new to the movement of thought. Only hypothesis formation points to the creativity that we have come to expect in thinking. Peirce is insistent: "No new truth can come from induction or from deduction."[54]

We should be careful in describing this "new truth." The creativity of hypothesis formation is in no way a creation ex nihilo or a creation from fancy (*phantasia*). Hypotheses are always situated in a particular context and proceed from past experience. Despite their situated character, the provisional conclusions that hypotheses draw provide new ways of thinking. In its ability to infer analogous conclusions from a variety of sets of predicates, hypothesis allows us to negotiate a future that unfolds in unusual and often unexpected ways. Indeed, hypotheses allow us to interpret the novelty of our surroundings not in terms of the challenges they present but rather in terms of the opportunities they afford. In this light, it will be worth discussing the family resemblance between the creativity involved in hypothesis formation and the aesthetic creativity posited earlier.

Peirce divides the realm of thought into hypothesis, induction, and deduction. He acknowledges, however, that these distinctions are artificial in an important respect. Hypothesis is *always* operative in conjunction with induction and deduction. In his description of Peirce's philosophy of science, A. J. Ayer highlights this point by saying, "to echo Kant [and Peirce], induction without abduction is blind [and] . . . abduction without induction is empty."[55] This statement, when couched in terms of a theory of the imagination, amounts to the claim that the imagination is always operative in the insight and fullness of human cognition. It is operative, yet often overshadowed, when thought takes an analytic turn. By the same token, deduction is also always present in the sense that it sets the stage for any future act of synthetic inference. For Peirce, there are no hard and fast delineations between analytic and synthetic modes of inquiry. Likewise, the domains of hypothesis, induction, and deduction seem to overlap and bleed into one another in the flow of inquiry. This said, Peirce believes that these realms can be described as theoretically and interestingly distinct.

Peirce's work on the nature of hypothesis is the first step in explaining abduction and musement, the ground and motivation of scientific creativity. As Anderson notes, this development does not get off the ground until the turn of the century.[56] It seems necessary therefore to move ahead in our chronological reading in order to trace Peirce's mature work on hypothetical inquiry. It is in abduction and musement that Peirce's epistemology most clearly reflects an imaginative disposition. Later it will be important to backtrack chronologically, picking up Peirce's attempts to apply his philosophy of mind in the construction of a pragmatic cosmology.

FOUR

ABDUCTION:
INFERENCE AND INSTINCT

Abduction: The Logic of Genius

Why were so many of Peirce's college days spent—some might say wasted—on the topic of genius? A look through his unpublished papers points to an obsession with the work and lives of those "great men" of extraordinary mental powers—from Michelangelo, to Mozart, to Edgar Allan Poe. Like his contemporary Josiah Royce, Peirce was fascinated by history and, more particularly, by the history of genius. This fascination might be attributed to Peirce's arrogant but not inaccurate suspicion that he would some day join the ranks of these "great men." I would suggest that Peirce's fascination relates to a more philosophical concern as well: his interest in the nature of human creativity. He was interested in the particular creative processes by which history advanced. In an early reflection on the nature of the creative imagination, Peirce insists that genius should not be restricted to the narrow confines of fine artistry. We should, he says, look elsewhere for the signs of imaginative genius: "The limitations of genius to art has met with universal disappointment among those authors who have written upon the subject. Jurgen Banalleyer expressly declares that there is such a thing as scientific genius."[1] The power of genius lies in its ability to generate novel solutions and hypotheses in the midst of human inquiry. It is not a power that can be restricted to the small arena of museums and concert halls. Peirce's interest in genius, pursued in cursory readings of Gerard and Schiller, quickly morphed into a exhaustive and exhausting analysis of hypothesis formation, the guesswork that underpins all creative activity. This analysis is known as Peirce's logic of abduction.

I would be remiss in neglecting a series of difficult questions that most Peirce scholars regard as a kind of rite of passage or, more accurately in some cases, a kind of trial by fire. What is abduction? Can abduction be formalized? How has abduction been described in the secondary literature? In light of this formalization and these more recent descriptions, what standards might be used to judge the verity and value of particular abductive processes? Most importantly for the purposes of the current project, to what extent does abductive logic coincide with the process of the imagination as described in this project? This last question is addressed in detail in the next chapter on musement and genius. But all these considerations shed light on Peirce's corpus and more generally on the nature of human cognition. In this respect, Jaakko Hintikka suggests that the questions concerning abduction stand as "the fundamental problem of contemporary epistemology."[2]

Abduction becomes a problem, perhaps "the fundamental problem," of contemporary philosophy of mind when it is viewed in comparison with inductive and deductive methods of inference. In light of the deficiencies of these two methods, the need for abductive logic comes home to us: abduction serves unique functions in the movement of thought, functions that neither induction nor deduction can perform. The distinctions that make abduction uniquely helpful also make it uniquely difficult to formalize. Since abduction departs from the methods of deduction and induction, it cannot be formalized in the terms of syllogistic or quantified logic. Indeed, on these grounds, some scholars suggest that there is no such thing as a logic of abduction or that at a minimum Peirce was helplessly confused when it came to articulating an abductive logic.[3]

I will argue that abduction expands the scope of logic. Once this scope is broadened beyond the deductive-inductive dyad that has traditionally characterized it, abduction reveals itself as logical. Not coincidently, a similar situation obtains between the imagination and the faculties of understanding and sense. Imagination performs epistemological functions that neither understanding nor sense can accomplish. It has been shown that the imagination compensates for and mediates the constraints of these two faculties. The process of imagination serves a vital role in the movement of thought, but since this role departs from the constraints of understanding and sense, it cannot then have recourse to these formal constraints in its description. Imagination and abduction remain exactly as necessary as they are ambiguous. This section strives to give a clearer articulation of both without downplaying the unique character of their roles in human reasoning.

Abduction can only be understood against a background of traditional modes of logic. Its characterization is therefore best developed in relief. To draw out the comparison between abduction and other modes of logical inference, I continue to trace the evolution of the concept of abduction in Peirce's work. The exchange in the late 1980s between Anderson and Robert Roth concerning the nature of Peircean abduction serves as an appropriate point of departure. This exchange revived the idea that Peircean abduction ought to be regarded as a type of instinct that cannot be reduced to mechanical processes and, perhaps more importantly, began to situate abduction in the context of Peirce's evolutionary philosophy. I highlight several more recent commentaries that elucidate the distinctive character of abduction and begin to propose formal models in its description. In the process, I argue that abduction is an ampliative logic better understood in terms of the strategic rules of modern game theory and artificial intelligence than by the explicative rules of deductive and inductive processes.

Although Peirce toyed with the concept of hypothesis formation as early as the 1870s and 1880s, it seems he only fully recognized the import of abduction at the turn of the century. In 1898 and again in 1902, Peirce explicitly identified the historical root of abduction as the Aristotelian notion of *apogogue*. At this point, Peirce suggests that *apogogue* refers to the process by which "the minor premise of . . . a syllogism . . . is inferred from its other two propositions as data." Aristotle inverses the deductive working of the syllogism, termed *epogogue*, in which a conclusion follows necessarily from given premises. What results in this inversion is the process of provisional hypothesis formation. Josephson and others structure their formal models of abduction around the belief that it is a formal reversal of deduction.[4] As Anderson notes, this definition of abduction was given in a somewhat abridged form in 1878 when Peirce lectured on Aristotle's treatment of propositional logic. He demonstrates this inversion by giving a particular fourth figure deductive syllogism:

Rule—All the beans from this bag are white.

Case—These beans are from this bag.

Result—These beans are white.

In this case, Peirce states that the corresponding abductive process would take the following form:

Rule—All the beans from this bag are white.

Result—These beans are white.

Case—These beans are from this bag.[5]

I will include the inductive form of the syllogism in order to lay the groundwork for the subsequent discussion of logic as a normative science of inquiry:

Case—These beans are from this bag.

Result—These beans are white.

Rule—All beans in this bag are white.

Peirce envisions that each of these three modes of inference performs a particular function in the development of inquiry. The deductive form is the only one of the three to possess apodictic certainty. Induction secures a type of truth, but this type is really only truth in the "long run" or what has already been referred to as the truth of habit.

The argument in the abductive case is not necessary in the apodictic sense—nor does it assure truth in the long run—but must be considered either probable or merely possible. Anderson echoes this point, writing, "in abduction the acceptance of the minor premise and of the entire syllo-gism is *provisional*."[6] Neither Aristotle nor Peirce suggest that this provisional type of reasoning be included in theory of the deductive-inductive syllogism but instead that it be considered another type of reasoning. Peirce quickly recognizes that this type of reasoning cannot be accommodated by a narrow formal context.

Describing abduction in terms of an inversion of Aristotelian syllogism will misconstrue this logical process unless we reconsider what Peirce, following in the footsteps of Aristotle, takes to be the root of logic. Logic is not the mechanical workings of a closed system but is the matter and working of open human inquiry. Anderson claims, I think rightly, that Peirce's lectures in 1898 suggest that "*apogogue* is both a logical form and a lived process."[7] With this suggestion, Peirce treads a precarious middle ground between logical formalism and an ampliative adaptive process. On these grounds, he will insist that abduction is both formally logical and procedurally ampliative. More simply, he claims that abduction is an ampliative logic.

Ampliative logic can be grasped negatively in the sense of being understood as *not* being explicative or deductive. In any deductively valid argument, there is a sense in which the conclusion is contained in the premises. Deductive reasoning serves the purpose of extracting information from the premises. In an ampliative argument, the conclusion "goes beyond" the premises by generating a new hypothesis that might help explain a given practical situation. In fact, the conclusions of *valid* ampliative inferences can be false even if their premises are true. Peirce reworks the syllogistic form to account for this ampliative character, describing the abduction as emerging in the form of a novel reaction to a surprising or problematic fact:

(F1) The surprising fact, C, is observed.

But if A were true, C would be a matter of course.

Hence, there is reason to suspect that A is true.[8]

In the preceding section, I discussed the fact that abduction, as the process of hypothesis formation, provides the suspicion of novel explanations. This point demands greater attention.

In Peirce's reformulation of the syllogism, it becomes clear that he envisions abduction as the beginning of or, more accurately, *as one recursive moment in* the logic of scientific inquiry. As Peirce comes to see by 1908, abduction has been neglected as a logical argument because of its weakness. But the neglect of abduction has resulted in a failure to describe accurately the spontaneous movement and growth of human inquiry.[9] Abduction, despite its relative weakness as a logical argument, stands as the logical form/process upon which the *movement* of science depends. Abduction provides a "reason to suspect" and, consequently, a reason to test and theorize. By its hopeful suggestion, abduction keeps inquiry on the move but at the same time constrains future moments of investigation. This will become clear when we examine induction in terms of Peirce's understanding of existential graphs in the next chapter. The "hence" or "therefore" of the (F1) form carries a pragmatic and normative weight.

Thus construed, abduction, according to Peirce, "recommends a course of action." While it is true that abduction "commits us to nothing," it is equally true that abduction causes a hypothesis to "be put down in our docket of cases to be tried."[10] As Kapitan notes, Peirce's language suggests that abduction creates the conditions for the possibility of novel choices, choices that are to be decided upon and to be "tried" by the agent of inquiry. He also suggests that the abductive creation of certain hypotheses and the neglect of others is already a type of choice, a type of "probational adoption," that tentatively recommends a course of normative action.[11]

As Roth and Anderson both note, it is in this sense that abduction fits into Peirce's understanding of normative science as evolving through a self-controlled and self-generating evolution. Abduction "is the first level of self-controlled reasoning and it therefore, on Peirce's view, ought to be the foundation of scientific inquiry."[12] At first glance, this claim seems a bit far-fetched. How is abduction self-controlled? After all, the abductive moment seems by definition out of our control. In Peirce's language, the "abductive suggestion comes to us as a flash."[13] This moment of insight corresponds to the working of genius, processes understood to happen by natural instinct or chance.

Peirce, however, is careful not to render abduction as a type of random perception or as just an instant of tychism, normally and inaccurately construed as the principle of arbitrary chance or pure contingency. Instead he holds that abductive processes reflect *both* instinctual insight and a type of logical inference. Harold Frankfurt in an early commentary makes this point by stating that Peirce holds the paradoxical position that "hypotheses are the products of a wonderful imaginative faculty in man *and* they are products of a certain sort of logical inference."[14] Later in his work, Frankfurt strays from his original position, setting imaginative insight against logical inference, a dichotomy that Peirce himself seems to draw into question. For Peirce there is an intentional conflation between insight, a type of spontaneous perception, and abductive inference. This conflation should be a familiar one. Kant makes a similar move to describe imaginative insight-judgment.

In "Pragmatism as the Logic of Abduction" (1903), Peirce reinterprets the Aristotelian claim—*Nihil est in intellectu quod non prius fuerit in sensu*—stating that nothing is in conceptual meaning that is not first in perceptual judgment. He goes on to say that perception provides the "starting point or first premise of all critical and controlled thinking."[15] Such insightful perceptions, however, are "results of a process, although a process not sufficiently conscious as to be controlled, or to state it more truly, are not controllable and therefore not fully conscious."[16] Immediately following this statement, he comments that abductive inference "shades into" these uncontrolled perceptual states "without any sharp line of demarcation between them." He elaborates on this position in 1911 in a short unpublished fragment: "But the one exceptional feature of [abduction] that forbids my calling it reasoning at all is that its result is forced upon us without any appeal for our approbation and is thus an automatic act or rather performance of our organism and not our free act."[17] Here, Peirce seems to rescind his earlier stance that abduction ought to be regarded as a type of reasoned inference to the extent that it remains a "performance of our organism." We should not, however, regard this unpublished manuscript as Peirce's final word on the subject. It is instead an indication that in his development of abduction, Peirce continued to think through the relationship between embodiment and reasoning, between instinctual insight and logical inference. I believe we are to take abduction as an operation on the cusp of, in the borderland between, nature and reason. Peirce has made two significant suggestions in this regard: (1) that logical inference, in its relation to the process of perceptual judgment, is in *a certain sense* out of our control and (2) that perception, in its relation to abduction, is already open to the purposive interpretations of logic. To put the latter point somewhat differently, perception and abductive insight, insofar as they arise from the ordered context of a particular embodiment, reflect a unique type of logical structure.

The Standards of Abductive Insight

While abduction blends with a type of spontaneous instinct, an almost miraculous insight, Peirce does outline principles by which an abductive hypothesis can be judged. Indeed, these principles "separate" abduction from the usual renderings of "mere" perception. According to Peirce, hypotheses must meet three interlocking standards. They must be explanatory, parsimonious, and testable.

Initially, it might seem that the first of these qualifications is the most important. To say that a hypothesis is explanatory, however, verges on tautology. As in the above case, a hypothesis accounts for the facts of the matter. This first criterion therefore gives hypothesis formation little direction; the injunction to be explanatory does not *narrow* the field of hypotheses. For example, the "surprising fact C" could be the consequent of *any number* of antecedent situations that could be described as event A. It is not any explanation that is required of abduction, however, but an *economic* one.

In light of the shortcomings of the first explanatory standard, the criterion of parsimony takes center stage. Peirce's appeal to the economy of hypotheses is reflected in his recognition that

> trillions of trillions of hypotheses might be made of which one only is true; and yet after two or three—or at the very most a dozen guesses, the physicist hits pretty nearly on the correct hypothesis. By chance he would not have been likely to do so in the whole time that has elapsed since the earth was solidified.[18]

Kapitan echoes Achinstein in the suggestion that there are a myriad of wild suggestions that might explain the surprising phenomenon, but it is the goal of abductive processes to weed out the wild guesses in order to hit upon a suggestion that is more economical than its "*envisioned* competitors."[19] It is for this reason that Kapitan revises Peirce's model given in (F1). It yields to:

(F2) Some surprising fact C is observed.

If H were true, then C would be a matter of course.

H is more economical that its *envisioned* competitors.

Hence, H is more *plausible* than its competitors.

Here the term "envision" seems significant to our study, for it resonates with a type of artistic imagining that has been addressed in earlier sections. Abduction requires a type of envisioning process, a type of insight to imagine the possible competing hypotheses and to determine the suitedness of a given suggestion. Once again, abduction appears two-faced, in the sense of comprising inference as well as insight. The suitedness of a given hypothesis may *seem* to be actively determined by an agent of inquiry, but it is also actively determined by the environmental and contextual factors in which the agent finds its being-in-the-world. Indeed, it may be more accurate—but perhaps more complicated—to say that the standards of a particular good guess are continually established in the nexus between an individual and a wider community.

It is worth noting that the suitedness of a hypothesis is measured in terms of the consequences of assuming the hypothesis as a mode of action. In the process of envisioning and selecting hypotheses, an organism also attempts to foresee the consequences of the provisional and hypothetical explanation in light of an overall strategy for a given inquiry or game. In terms that will be employed in later sections, the suitedness of abduction is a function of organism fitness, not only in a sense restricted to its evolutionary connotation, referring to the ability to produce offspring, but understood more generally as the ability to achieve harmony with particular environmental conditions.

The suitedness of abduction points toward both the ontological underpinnings on which all hypothesis formation depends as well as toward abduction's transcendental underpinnings. It points to the fact that the abduction that animates our mental life is *intimately* constrained and directed by material factors that have traditionally been regarded as ontologically distinct from agents of inquiry. This point is made explicitly in Peirce's observations in regard to the conditions for possibility of abduction.

> Nature is a far vaster and less clearly arranged repertory of facts than a census report; and if men had not come to it with special aptitudes for guessing right, it may well be doubted whether in ten or twenty thousand years that they may have existed their greatest mind would have attained the amount of knowledge that is actually possessed by the lowest idiot. But in point of fact, not man merely, but all animals derive by inheritance two classes of ideas which adapt them to the environment. In the first place, they all have from birth some notions, however crude and concrete, of force, matter, space and time; and in the next place, they have some notion of what sort of objects their fellow-beings are, and how they will act on given occasions.[20]

Here Peirce hints at an ontological and evolutionary explanation of why the human intellect, largely defined by its abductive ability, is in his words "peculiarly adapted to the comprehension of the laws and facts of nature." In the late 1960s, W. V. O. Quine echoed this point, stating in "Epistemology Naturalized" that to trust induction and, by extension, other nonmonotonic procedures such as abduction, "as a way of access to the truths of nature . . . is to suppose . . . that our quality space matches that of the cosmos."[21] This point will be developed in the coming chapters as we investigate the bodily basis of abductive processes.

For the moment, however, we reexamine the Janus-faced character of abduction as being both inference and insight while being mindful of the ontological implications this reexamination entails. Peirce describes this process of envisioning as an "act of instinctive insight,"[22] as being "neither more or less than guessing."[23] In the same breath, he insists that this insight operates, in some respect, by way of reasonable judgment and in this regard should be considered a true form of logical inference. As Kapitan writes, "Hypotheses are not generated fortuitously. [Peirce's] remarks point to the inferential nature of guessing, implying it is under our control."[24] Kapitan takes this to mean that abduction is under our conscious control. This is only partially correct. Abduction is simultaneously under the control of natural and environmental tendencies that enable "our" abductive guesswork. This twofold character was noted in Kant's rendering of imaginative genius, which is both actively enacted by an individual and bestowed as a type of natural gift. The creation of genius depends on an attunement or receptivity to the unfolding of nature and the active disposition that "hits upon ways of expressing" this unfolding. Abduction—as insight and inference—resonates with the dual character of genius.

It is important to pay close attention to the words that Peirce uses to describe the parsimony or economy of suited hypotheses. "Before you try a complicated hypothesis, you should make quite sure that no simplification of it will explain the facts equally well."[25] It seems to follow from the earlier comment on the "envisioning" of abduction that one is to imagine/see the possible alternatives to a given hypothesis and, in the same process, determine the hypothesis that is a "good bet" according to this criteria of economy. Since all forms of explanation are fitted to various specific contexts and serve various purposes, the most "economical" form of a hypothesis will vary depending on the particular purposes of investigation. In his description of parsimony, Peirce also comments that suggestions must be "natural" and "likely,"[26] must reflect family resemblance with familiar knowledge, and must be cost-effective in their testability. In sum, chosen hypotheses must be deemed naturally expedient regarding a possible future end and naturally fitted regarding the body of past experiences and judgments.

While Peirce thus supplies a type of rule for abductive processes, economy and "natural fittedness" are not the type of rules that produce the closed system of deductive logic. The standards that govern abduction are general, open-ended principles rather than defined rules that prescribe an exact course of action on a case-by-case basis. As Jaakko Hintikka notes, this is the distinction between strategic and definitory rules. "Definitory rules are merely permissive. They tell us what moves one may make in a given circumstance, but they do not tell anything about which moves are good, bad or indifferent."[27] Hintikka also describes the standards for validity in the case of definitory rules, noting that "they are validated in so far as they confer truth or high probability to the conclusion of each particular application of theirs."[28] In contrast, strategic rules are justified to the degree that they eventually lead to strategies that pursue the purposes or ends of an investigator. It is important to note that this end is never fully in view and, more often than not, is fully shielded from view. Holding the question of ends in abeyance, however, it is possible to say that strategic rules cannot be judged on a case-by-case basis but only in the context of a complete strategy. An analogy can be found in game-theoretical models when general utilities are calculated not in terms of a single move but over a series of iterated moves, in other words, in terms of a complete strategy. In his study of ampliative logic, Fredrick Wills draws out this distinction between deductive and abductive game playing and puzzle solving. "Not all the pieces in [the ampliative] 'puzzle' are given antecedent to the process [of solving it]. Rather it is only when this kind of puzzle is solved, and as a consequence of the solution, that what constitutes the pieces and the nature of the pieces, are for this puzzle finally determined."[29]

The strategic calculations made in abductive puzzles are not made on the basis of probability, that is, by way of induction, but on the basis of an ability to see and anticipate the way that a possible hypothesis might *plausibly* contribute to a future harmonious strategy. Abduction provides a heuristic way of understanding an array of possible hypotheses, a type of intellectual/instinctual shorthand that makes a provisional judgment concerning this multitude of possible explanations. Again, we see the inherent fallibility of abductive conclusions. These conclusions depend not only on a type of *in*sight—what Peirce will occasionally call the natural light or *lume naturale*—but on a type of *fore*sight that can see beyond the premises of a given situation in order to generate hypotheses. These hypotheses lead an inquirer to unforeseen truths when pursued as a long-term policy.

The foresight of abduction that guides human inquiry, however, is limited in two significant respects. First, our foresight is often painfully shortsighted. As Peirce points out, we cannot imagine some strategies in their entirety. Second, the actions that abductive hypotheses recommend sometimes run counter to past strategies and may be aborted on these grounds before their long-term benefits can be realized. Both of these difficulties boil down to the fact that while we may possess the insight and foresight to make educated guesses in light of future possibilities, we lack the *perfect* insight that would be required to understand these future possibilities as actual matters of fact. In effect, we do not know what the future holds but are left to our educated and imaginative guesswork.

Hintikka suggests that the abductive process might be understood in the context of a broader game-theoretical strategy. However, he also claims that Peirce failed to incorporate the notion of strategy into his logic or otherwise put it to use. This criticism seems to neglect Peirce's understanding of inquiry as prescribing modes of action that are in some way directed toward possible optimal outcomes. Peirce's rendering of habit as the assumption of patterned regularities over time seems in line with this thesis.[30] Habit, however, while providing the basis for consistent reactions to given stimuli, cannot account for an organism's adaptive behavior in the face of surprising facts and situations. Only abductive processes can supply these imaginative adaptations. Before returning to a direct discussion of the imagination and aesthetic genius it may be worthwhile to propose a sketch of a formal strategic system that might accommodate Peircean abduction. This sketch may give us a better handle on the character of abduction, even as it allows us to explore the extent to which the imagination can be described in formal or computational terms.

PEIRCE: The Game of Abduction and the Imagination

Since the 1960s, practitioners in game theory and artificial intelligence have understood the crucial role that abduction plays in question-answer logics and strategic systems. Formalizing its role, however, has proven extremely difficult. Here I briefly describe one of the most recent attempts and highlight the difficulty this model faces in resolving the boundary conditions and preference criteria for possible hypotheses. In light of this difficulty in making abduction computational, we are forced to turn our attention to Peirce's later work, which sheds more light on the imaginative character of abduction.

In the late 1990s, researchers began to develop a computational model that would simulate the process of abduction. This "abductive tool," aptly named PEIRCE, was created "to enable knowledge engineers to determine when a particular explanatory method is appropriate for invocation and how the process selects a method for invocation."[31] At first blush, this description seems to grasp the heart of abduction, namely, the goal of selecting hypotheses. Ultimately, however, it is clear that PEIRCE fails to master the defining ability of abduction: the ability to generate genuinely novel hypotheses in light of environmental circumstances. PEIRCE's failure is instructive in several significant respects. Most pointedly, it suggests that computational systems that have adeptly modeled inductive and deductive systems are inadequate when it comes to simulating the process of abduction. The weaknesses of PEIRCE help identify aspects of hypothesis formation more satisfactorily addressed by an imaginative approach.

The task-subtask description of PEIRCE reflects the common understanding that the primary goal of abduction is to explain the facts—surprising or otherwise—that are given. The top-level task of fulfilling this goal is divided into three subgoals:

> Obtain a candidate set of hypotheses for possible inclusion in the compound explanation that explains a wider field of findings. This initial set of hypotheses stands as the field of all possible explanations and is limited by a hierarchy classifier that organizes the possible explanations into groups according to preset/enumerated boundary conditions.
>
> Explain the findings by constructing the most plausible explanation that does not contradict compound explanation.
>
> Critique this compound explanation.

As Josephson and others note, these subgoals "provide a potential solution to the top-level goal, not a necessary one." That is, we satisfy the primary objective through the serial or mixed use of *any* combination of these subgoals—*or by using completely different subgoals.*

Without ignoring this caveat, we can quickly place each of these subgoals in the form of a general algorithm. I have omitted the issue of parameters and boundary conditions for the time being. The PEIRCE algorithm can be narrowed in the following manner:

1. Select a single finding that needs to be explained.
2. From the set of hypotheses found by the hierarchy classifier, select those hypotheses that offer to explain findings and that are not initially ruled out.

3.   If only one hypothesis is offered to explain the finding, select that explanation. Otherwise, select the most plausible explanation of the set.

4.   Integrate this hypothesis into the compound explanation. If the introduced hypothesis is incompatible with the existing compound (an incompatibility is a pairwise exclusive relation between hypotheses), two choices exist:

    a.   Select another hypothesis and return to step 3, or

    b.   Remove incompatible hypotheses in the compound explanation and unmark/reevaluate the findings that these now-discarded hypotheses explained. Maintain a history mechanism to avoid the loops.

5.   Given integration in step 4, update findings that integration might explain.

6.   If findings still remain unexplained, go to step 1. Loop.

Josephson and other researchers in artificial intelligence and computability have provided a computable model, including boundary conditions, for the general abductive algorithm.[32] All of the details of this computational model, however, are not necessary to evaluate PEIRCE in order to demonstrate the extent to which this formalization can account for Peirce's notion of abduction. The evaluation is lukewarm at best.

Problems for PEIRCE arise immediately, as soon as it becomes clear that this computational model depends on a "hierarchy classifier" (set out in step 2) to obtain a set of possible hypotheses over a given domain. According to the mature Peirce, this mode of selection is never given prior to abduction. Indeed, the work of the classifier, in its ability to generate a set of possible hypotheses, *is* the insightful activity abduction. Generation of hypotheses—not the adoption of them—is the true work of abduction. This work is never done beforehand, but in the midst of an interactive moment between an agent of inquiry and the world. It is the actual and unique embodiment of inquiry that places the initial—and fairly loose—constraints on the range of possible hypotheses. One is never able to give an exhaustive enumeration of these hypotheses, and only in rare cases is an inquirer vaguely aware or conscious of the constraints that limit the field of possibility. In the end, PEIRCE regards abduction as a mere "evidencing process." Fann and Burke both note that Peirce begins to distance himself from this stance as early as the mid-1870s.[33]

A second shortcoming of the PEIRCE model arises in the fourth condition (4). It becomes clear at this juncture that the model's attempt to simulate the criterion of economy is out of kilter with Peirce's description of abductive parsimony. In the model, each element in the set of possible hypotheses is given a "confidence score" that can be computed as a particular probability or in another confidence vocabulary. This, however, amounts to presupposing that a given confidence level can be designated prior to the engagement of abduction. This presupposition is the very thing that Peirce wants to explain in and as the process of abduction. Hypothesis formation relies on a type of educated guesswork—not to be confused with the calculation of relative frequency probability—in order to discover a suitable strategy to handle novel circumstances. As Peirce himself says in his Cambridge lectures of 1898, "the difficulty is that in most cases where resort is to retroduction [abduction] you know nothing about factual probabilities."[34]

A final comment needs to be voiced on the character of the model presented above. In the fourth constraint on the abductive assembly task (4), we are instructed as to how one might chose between hypotheses that explain the same observed event. Again, this decision is not based on insight or a natural receptivity but on two determinate parameters—the first being the confidence scores assigned to the hypotheses and the second a preset threshold used to measure a "significant" difference between these hypotheses. In the process of human investigation, no such threshold can be described in any discrete sense. This is not to say that there are not decisive moments in inquiry. There certainly are. Inquiry is full of moments in which we "come to our senses" and propose a possible solution to a puzzling situation. Rather, I suggest that the reasons for choosing one hypothesis over another cannot be given in a determinate concept or even anticipated in a type of pattern recognition. Peirce himself anticipates the difficulty in developing computational approaches to modeling cognition. In 1887, he reflects on logical machines that bear a remarkable similarity to PEIRCE. "Every reasoning machine, that is to say, every machine, has two inherent impotencies. In the first place, it is destitute of all originality and of all initiative. It cannot find its own problems; it cannot feed itself. It cannot direct itself between different possible procedures."[35]

The questions of initiative, purpose, insight, and originality that lie at the heart of abduction must, according to Peirce, "be thrown upon the mind" if we are to describe the activity of thought in a faithful manner. In other words, it is precisely these "impotencies" of machines that must be examined in any comprehensive investigation of human cognition.

Three critical points on the character of computational models of abduction emerge from this consideration of PEIRCE. In truth, however, these points trade on a single issue: the PEIRCE setup conceives abduction in only *one* of the ways that Peirce describes it. Since the 1960s, formal models invariably interpret abduction as an "inference to best explanation."[36] As Lorenzo Magnani notes, this interpretation is overly narrow. He suggests that abduction should be understood as bearing at least two valences. First, he agrees with scholars such as the Josephsons and Gilbert Harman: abduction should be considered a type of logical inference to the best explanation. However, he puts forth a second necessary interpretation of abduction that must accompany the first. Abduction must maintain its character as a creative process of generating novel hypotheses.[37] Once we acknowledge that a model like PEIRCE cannot represent the creative side of abduction, its shortcomings seem less destructive. The formalization in PEIRCE, if it does not succeed in portraying Peircean abduction, does succeed in exposing the limits of particular computational models and encourages us in the pursuit of more nuanced formal models. I develop such a model that employs Peirce's later work on existential graphs at the end of the next chapter.

If nothing else, the investigation of PEIRCE throws us back on the character of creativity. In so doing, it encourages us to return to the discussion of the creative imagination. It is worthwhile to explore the extent to which Peirce's later forays into epistemology hide a robust theory of the imagination, one that might begin to explain aspects of abduction that have so far befuddled researchers in AI. The movement and processes of the imagination may shed light on aspects of abduction that have been hitherto overlooked.

Cautionary Words Concerning Creative Instincts

My approach in the coming chapters will investigate abduction as a type of creative process that cannot simply be reduced to inference to best explanation. Abduction is not simply the process, proposed by PEIRCE, of determining which given "explanatory method is appropriate for invocation" but rather is the imaginative insight to see *new* explanatory frameworks for confusing phenomena. I am admittedly interested in abduction as a type of instinct or natural affordance, but I also realize the danger in pursuing this interest too far. Genius has often been regarded as a mysterious power that is bestowed upon a few select individuals. Genius is a "natural gift." Some people get it. Others do not. Simple as that. The supposedly occult nature of genius has obstructed investigations into the sociological, economic, and cultural factors that have led certain individuals to be regarded as geniuses. Similarly, the ineffable character of the imagination has created the unquestioned divide between genuinely creative animals (historically regarded as "humans") and mere brutes. Imagination is a natural gift. Some species get it. Others do not. Simple as that. *It is not as simple as that.* To suggest that the creative imagination is a natural gift is not to give up on its explanation but rather to suggest that an adequate explanation of creative imagination might be far harder than we originally thought. Such an explanation would not merely look at the creative powers of specific human beings working away in particular cultural settings but rather would look more carefully at the natural preconditions that structure this creativity. For Peirce, the question of creativity or abduction is not primarily one of human aesthetics or epistemology; it is an ontological query that demands that we trace out the processes of the imagination, its mediation and spontaneous emergence, in the workings of nature.

FIVE

IMAGING NATURE

In the aesthetic state everything—even the tool that serves it—is a free citizen, having equal rights with the noblest; and the mind, which would force the patient mass of the body beneath the yoke of its purposes, must first attain its assent.

—Friedrich Schiller, *Letters on the Aesthetic Education of Man*

"The Art of Reasoning": Genius of Imaginative Inquiry

In the spring of 1887, having been unceremoniously discharged from his post at Johns Hopkins University, Peirce developed a correspondence course on logic and critical thinking.[1] The course, on the fundamentals of Peircean logic, was aptly titled "The Art of Reasoning." During this period, Peirce wrote a series of letters that began to explain the imaginative-abductive basis of logic and cognition, and his early observation that "poets see a common nature" came to fruition. Peirce continued his attempt to describe reasoning as a form of art over the next years. By the turn of the century, the topic moved to the fore in his writing.

On December 23, 1908, Peirce looked back on his earlier work in a letter to Lady Welby. In it, he repeatedly refers to the works and writers that had inspired his interest in abductive reasoning, the disposition of "musement," and the "play" that accompanied it. "As for the word play, the first ever read . . . was Schiller's *Aesthetic Briefe* where he has so much to say about the *Spiel-Trieb*; and it made such an impression upon me as to have soaked my notion of play to this day."[2] As mentioned earlier, Peirce's admiration for Schiller is deep and heartfelt. In his *General and Historical Survey of Logic*, Peirce recounts the many days he spent during his youth reading the text, even as he acknowledges the shortcomings of his immature reading. In one of these reflections, Peirce writes:

> I must thank my good angel, in the form of Anna Lowell, that with-out knowing the wisdom of what I was doing, the first twelve months of my studies of philosophy was devoted to the reading, rereading, pondering, and repressing of Schiller's *Aesthetic Brief* by which I was at once introduced to Phenomenology, which ought to occupy the first year or two of every philosophy infant.[3]

Despite his enthusiasm for Schiller's aesthetics, Peirce abashedly ad-mits: "I read various works on esthetics, but on the whole, I must confess that, like most logicians, I have pondered the subject far too little. The books do seem so feeble. That affords one an excuse."[4] The feebleness of the books, however, is not the principal cause of Peirce's neglect. He continues to explain that "esthetics and logic seem, at first blush, to be-long to different universes." In his "Logic of Relatives," first developed in the early 1870s and elaborated in the late 1890s, Peirce begins to mediate between these opposing disciplines:

> It is only very recently that I have become persuaded that this seem-ing is illusory, and that, on the contrary, logic needs the help of es-thetics. The matter is not yet clear to me; so unless some great light should fall upon me before I reach that chapter, it will be a short one filled with doubts and queries mainly.[5]

It is not surprising that Peirce draws heavily on Schiller in his develop-ment of normative science, abductive play, and triadic logic. By adopting the poet's description of *Spiel*, Peirce finds a type of phenomena that demonstrates the novelty, growth, and mediation of abductive processes. Schiller very obviously believes that it is through artistic and interactive *Spiel* that one comes to recognize the order of things as the ongoing or-dering of mind. Schiller's work grounds Peirce as he thinks through the relationship between abduction and natural processes. As Schiller writes in his *Kallias*, "Nothing is free in nature [in the sense of giving itself its own rule 'through reason'], but at the same time, nothing is completely arbitrary either."[6] The processes of nature hover between the rational and the arbitrary, in a field where the imagination has free play. Here we return to the disposition of freedom-within-constraint that grounded our earlier discussion of abduction. The concept serves as the point of departure in the upcoming discussion of Peircean *agape*.

What is surprising in the young Peirce's reading of Schiller is that he is seemingly unaware of the wellspring from which the poet received his philosophical nourishment. Schiller's notion of play derives in large part from Kant's work on the imagination. As Barnouw and others have noted, Schiller adopts Kant's understanding of feeling (*Empfindung*) in the development of his aesthetics and psychology.[7] In this regard, Schiller bypasses the *Critique of Pure Reason* and assumes the project of examining aesthetic judgment as given in the *Critique of Judgment*. While he arguably does a better job than Kant in describing artistic play and production, Schiller maintains the basic structure of reflective judgment, preserving its hypothetical and spontaneous growth as well as its mode of justification. In letter 18, it becomes clear that Schiller, like Kant, is in search of a process that might mediate between sensation and understanding—the process of aesthetic imagination. In letter 20, Schiller writes:

> The mind goes from sensation to thought through a middle disposition in which sensuousness and reason are active at the same time and for that very reason take away each other's determining force and bring about a negation by way of an opposition ... thus one must call this condition (attunement) of real and active determinability the aesthetic condition.[8]

Schiller identifies the bridging of sensation and reason as an aesthetic process. He then sets the stage for Peirce, who thoroughly integrates this insight into his epistemology.

Peirce's pragmaticism therefore extends Kantian imagination (via Schiller) in a pragmatic and relational inquiry.[9] Somewhat ironically, this inquiry is meant to expose the inadequacies of the first *Critique*, a work Peirce believes summarizes Kant's corpus on the whole. At this point, it should be obvious that such a summary is inadequate at best. By largely overlooking the *Critique of Judgment*, the very work that Peirce himself disregards, commentators such as Karl-Otto Apel are able to assert without qualification that "Peirce replaces Kant's alternative of synthetic *a priori* and synthetic *a posteriori* propositions with the fruitful circle of correlative propositions of hypothetical abductive inference and experimental confirmation."[10] Justus Buchler anticipates Apel's reading when in 1939 he comments that Peirce was right to dismiss Kant on the grounds of the German's interest in the *Ding an sich*. "Looked at from [the] point of view [of the thing-in-itself]," Buchler writes, "Kantianism is actually contrary to pragmatism, which, if it condemns anything, condemns the view that there are a priori synthetic propositions."[11] Several qualifications are in order. As developed above in chapter 2, Kant himself replaces the dichotomies of the first *Critique* with the fruitful circle of aesthetic play and reflective judgment. Through this replacement, he provides an imaginative legacy to Peirce, who either overlooks this move in Kant or downplays its significance. Reflection on Peircean musement and abduction elucidates this legacy. Musement, as Peirce describes it in the "Neglected Argument for the Reality of God," is a moment of Kantian reflexivity, imaginative play. Peirce fleshes out the concept in this way:

> It is pure play [having] no rules except this law of liberty. It bloweth where it listeth. It has no purpose, unless recreation. . . . It begins passively enough with drinking in the impression of some nook in one of the three Universes. But impression soon passes to attentive observation, observation into musing, musing into a lively give and take between self and self. If one's observations and reflections are allowed to specialize themselves too much, the Play will be converted to scientific study and cannot be pursued in odd half hours.[12]

Play's purpose is re-creation. Its purpose is literally to *create again*. The muser is attentive and, like the artistic genius already described, receptive to the natural ordering of being. Musing abduction "bloweth where it listeth." The voices of these verbs are intentionally ambiguous—hovering oddly between activity and passivity. One can blow but also be blown. To list is to desire but also to be compelled. For Kant, aesthetic ideas arise in the ground between desire and compulsion. For Peirce, this imaginative middle ground is the fertile plot where free reasoning comes to fruition.

Here, Peirce seems to know the precise connotation of his words. Grammatically, he is attempting to express the to and fro, the give and take of imaginative *Spiel*. Musement and hypothesis formation, like Kantian aesthetic judgment, is not prescribed by any a priori rule or constraint but rather discovers *and* develops the constraints of an evolving situation. Musement cannot be described in terms of traditional logical analysis. Musement cannot be logically described. It nonetheless gives rise to a specific type of inquiry.[13]

As we have seen, Peirce repeatedly comments that abductive reasoning has the structure of hypothesis formation and acts between the natures of deduction and induction. Early in his career, Peirce asserts:

> An abduction is a method of forming a general prediction without any positive assurance that it will succeed either in the special case or in the usual, its justification being that it is only a possible hope of regulating our future conduct rationally, and that Induction from past experience gives us strong encouragement to hope that it will be successful in the future.[14]

The certainty of abduction, like the certainty of reflective judgment, remains strictly provisional since it cannot rely on determinate rules or concepts. Just because hypothesis "gives us strong encouragement to hope that it will be successful in the future" does not mean that one can rely on this hope unconditionally. Again employing Schiller, Peirce writes that "Mr. Schiller himself seems sometimes to say, there is not the smallest scintilla of logical justification for any assertion that a given sort of result will, as a matter of fact, either *always* or *never* come to pass."[15] It is interesting to note that Schiller, a poet and artist, inspired Peirce to think through the limits of logical quantification—the fact that events do not occur with absolute regularity. Peirce, by way of Schiller, is recapitulating Kant's understanding that the play of the imagination can neither be predicted nor determined—in the affirmative or the negative—for all time.

At the beginning of this book, I discussed the way in which Kant's genius was unable to provide an articulate account of his or her aesthetic apprehension. The apprehension outstrips any attempt to describe it. The aesthetic apprehension is, in a rather odd sense, not in the hands of the genius but is instead offered up as an *ingenium*, as a gift of nature. It seems necessary to return to this point to highlight the imaginative character of Peircean abduction. Peirce's scientific inquirer cannot "give reason for abduction . . . and it needs no reason since it merely offers suggestions."[16] In a certain sense, Peircean inquiry is beyond the knowledge and the power of the inquirer. To use Kantian language that remains faithful to Peirce, the inquirer is the "natural gift" through which nature gives the rule to investigation. Abductive play is the "performance of our organism." The inquirer is at once productive and receptive in reference to the rule and performance of abduction. Again, this rule is *not* a concept but an informed and "original suggestion," a type of natural prompting. Peirce stresses this instinctual, giftlike quality of abduction when he writes in 1898:

> The third kind of reasoning [namely abductive reasoning] tries what *il lume naturale*, which lit the footsteps of Galileo, can do. It is really an appeal to instinct. Thus Reason, for all of the frills that it usually wears, in vital crises, comes down upon its marrow-bones to beg the succour of instinct.[17]

Abduction depends on one's ability to listen and respond to the natural ordering of the world. It is an embodied capacity to make inferences—a logic that is in the bones of an individual. This ability stems from the root of bodily instinct. It is instinct, and presumably the particular embodiment of an inquirer, that first constrains any given investigation.

Peirce elaborates on this necessary continuity between the modes of investigation and the natural world in "The Architecture of Theories."

> Thus it is that, our minds having been formed under the influence of phenomena governed by the laws of mechanics, certain conceptions entering into those laws become implanted in our minds, so that we readily guess what those laws are. Without such natural prompting, having to search blindly for a law which would suit the phenomena, our chance of finding it would be one in infinity.[18]

It is true that abductive reasoning simply amounts to a type of guesswork. But here Peirce notes that this guessing is not simply random. Abduction relies on the fact that nature lends itself to the orderability of mind and, indeed, that this order is of nature. It is also important to note that while abduction is provisional, it is also real and irreversible. Hypotheses change the landscape of future inquiry. This statement echoes the earlier discussion of the nature of the *ingenium* of Kantian genius. In one instance, Peirce even acknowledges this connection between abduction and imaginative genius. He suggests that the realities of nature compel us to put some things into very close relation and others less so. "But it is the genius of the mind, that takes up all these hints of sense, adds immensely to them, makes them precise, and shows them in intelligent form in the intuitions of space and time."[19] This is precisely the role that imagination assumes in reflective judgment as presented in the third *Critique*. Peirce, in the "Methods for Attaining Truth," like Kant in the *Critique of Judgment*, repeatedly insists that the genius of abductive inquiry must simultaneously be receptive to the "natural light" of the world's order and be continuous with this *lume naturale*. He argues that if the general observations of the universe demonstrate its conformity to a type of lawfulness, and if the human mind has been shaped under the force of these laws, then "it is to be expected that [human beings] should have a natural light, or light of nature, or instinctive insight, or genius, tending to make them guess those laws aright, or nearly aright."[20] Peirce insists that there is a type of free attunement, a kind of genius, that allows one to study the ordering of nature.

Pragmatic Common Sense and the Community of Inquiry

In surveying Peirce's expansive corpus, one is initially struck by his hesitation to speak of unconditional truth. This hesitation is understandable when one recognizes his total reliance on abductive reasoning, on the creative guesswork that sediments his notion of pragmatic justification. In a typically broad generalization, Peirce goes so far as to state that "if you carefully consider the question of pragmatism, you will see that it is nothing other than the question of the logic of abduction."[21]

A puzzlement, however, remains. How might the question of pragmatism, the question of abduction, be answered with any type of certainty? It seems quite obvious that Peirce does not want to retreat to the relativism that has come to define much of present-day pragmatism. He does not want to abandon certainty all together. In light of this project's comparison of the imagination and Peirce's thought, it seems appropriate to ask a series of questions that might lead us in the right direction. If abductive reasoning, in the Peircean sense, shares a family resemblance with Kant's notion of aesthetic (imaginative) judgment, might one also expect to find a similarity in the justificatory frameworks of each mode of inquiry? Might aesthetic "certainty," grounded in an imaginative process, and abductive "certainty" be of the same order? More specifically, might one expect to find a version of the Kantian *sensus communis*, as described in chapter 2, and an imaginative sensibility lurking beneath the surface of Peirce's formal system? It seems possible to answer in the affirmative in each of these cases.

It seems fairly straightforward to say that a particular abduction cannot be proven as formally and universally true. Instead, it is experienced in a type of felt harmony in a particular situation. Abduction is experienced as the initial call to action, as an initial proposition. Effective abductions are affective. The goodness of a good guess is felt—for the time being—in its active mediation of the circumstance. It is in this *sense* that Peirce remarks that "the mediating triad . . . has . . . for its principal element merely a certain analyzable quality *sui generis*. It makes (to be sure) a certain feeling in us."[22] This qualitative feeling, however, is not merely subjective. Rather it always maintains the possibility of communication and evaluation by a community of inquirers. Peirce is careful with his wording: "the mediating triad . . . makes a certain feeling in *us*." It is in this unique *common* sense that I will suggest that we interpret one of the justificatory norms of Peircean abduction as a type of pragmatic *sensus communis* that motivates and underpins Peirce's convergence theory of truth.

In "A Survey of Pragmatism," Peirce writes that Schiller motivated him to unify subjective feeling and objective constraint and encouraged him to develop a "conditional idealism." In developing his pragmatism, Peirce could neither ignore the pervasive affective quality of an abductive situation nor eschew the notion of a common sense that might provide the justificatory ground of human inquiry. The "conditional idealism" that results has already been partially described as an offshoot of the imagination in Kant's development of the aesthetic *sensus communis*, a sense that is paradoxically free and determinate. This conditional idealism depends upon its testability in a community of inquirers; this is the final criterion of abductive processes. For Kant, the communal sense is determinate insofar as it establishes the "rules" or, more appropriately, the tests for genius. It is free precisely to the extent that genius expands these rules, passes these tests, and extends the established guidelines of the *sensus communis*. The pragmatic common sense that Peirce inherits evolves and is evolving. Interestingly, the genealogical movement of thought presented in "The Law of Mind" and the continuous lineages highlighted in "Evolutionary Love" seem to progress in similar fashion, adhering to similar modes of justification. These articles are addressed in detail in the coming sections of this chapter in order to examine the cosmology and ontology that is implied in Peirce's imaginative philosophy of mind. For the moment, it seems appropriate to spend a bit more time describing the imaginative movement of thought that Peirce begins to sketch out.

For Peirce, the movement and justification of thought is affected by the entire history of thought, by all previous thoughts, to a greater or lesser extent. It is Peirce who inspires William James's comment in 1906 that the pragmatist holds his or her ideas true only to the extent that everyone freely consents to this true position in the long run. Bruce Kuklick underscores this similarity between Peirce and James, stating that both viewed absolute truth as an "inevitable regulative postulate."[23] In his 1890 "Logic and Spiritualism," Peirce writes that "common sense corrects itself, improves its conclusions. . . . [But] it does not then attain infallibility."[24] As David Savan remarks, "Peirce conceded that 'there may be a question that no amount of research can ever answer,' and that the real world is in some respects indeterminate."[25] Peirce's emphasis on the ineffable is what separates his "conditional idealism" from the absolute idealism of F. H. Bradley or the early Josiah Royce.

If truth is always measured and revised according to some environmental, historical, or social context, it seems to follow that abduction must be understood in a similar way. The suitedness of a particular abduction rests on the entire history of guesswork—both human guesswork and the evolutionary guesswork that arrived at the tentative conclusion of the human species. Being bound to this history does not mean that a particular thought is not free. It simply means that thought is both free and constrained. The genealogy of mind delimits the field of free possibility for future action and thought. It is clear that Peirce's notion of the mind's continuity does not only apply to the individual mind of an embodied human being but also to the collective Mind of the community of inquiry. One ought to remember that imagination plays between the subjective feeling of judgment and this feeling's continuity with the nature of the *sensus communis*. Josiah Royce (after a sustained correspondence with Peirce in 1902) makes this point more poignantly and with an emphasis on community and intersubjectivity. But Peirce also seems well aware of its implications. In an unpublished and undated manuscript entitled "The Theory of Reasoning," Peirce describes the relationship between the truth and the inquiry of an individual human being, outlining a conditional idealism that depends on a collective pursuit of the truth.

> This ultimate destiny of opinion is quite independent of how you, I, or any man may persist in thinking. It is thought, but it is not my thought or yours, but is the thought that will conquer. It is this that every student hopes for. It is the Truth; and the reality of this truth lies, not at all in its being thought, but in the compulsion with which every thinker will be made to bow to it, a compulsion which constitutes it to be exterior to his thought. If this hope is altogether vain, if there is no such compulsion, or externality, then there is no true Knowledge at all and reasoning is altogether idle. If the hope is destined only partially to be realized, then there is an approximate reality and truth, which is not exact.[26]

We hope that our individual inquiries are not made in vain. We hope that they are "on the right track." We hope to make some original contribution to the direction of this circuitous path. We act on the grounds of this hope and, in so doing, engage in the movement and evolution of Mind. The realization of continuity itself—continuity with the social nature of Mind—becomes the dynamic benchmark of Peircean "certainty." In the movement of common sense, an inquirer comes to feel a type of continuity in the lineage of Mind. Recognition of the continuity of Mind allows one to at once acknowledge and disclose the continuity of nature.

These final suggestions seem rather far-flung. At this point, it may seem that Peirce has abandoned his careful study of logic and epistemology in favor of a hopeful idealism or panpsychism. In the coming section, it will be necessary to show in detail how the process of creative imagination in which Peirce's epistemology and logic are grounded encourages him to reassess the character of the natural world. It encourages him to explore the continuity of nature and mind. It encourages him to construct a cosmology and an idealism that remain faithful to his commitments as an empirical scientist and logician who aims to describe accurately the character of human inquiry.

Abduction Replayed: Diagramming the Spiel of Inquiry

In this section, I review the ground covered and survey a largely uncharted area in Peirce scholarship in order to make room for the formalization of abductive processes. We have seen how computational models such as PEIRCE cannot account for the imaginative character of abduction. They tend to ignore the ampliative elements of hypothesis generation, downplay the way in which hypotheses arise from embodied and historically situated contexts, and overlook the unique standards that serve as the desiderata of Peircean abduction. We have also begun to outline the genealogy of abduction, tracing its origins to the imaginative aesthetics of Schiller and, more speculatively, of Kant. I propose an alternative model to PEIRCE that describes abduction as a vital part of a query-driven process, a process that resonates with the reflective process of aesthetic judgments described earlier. More significantly, and with more difficulty, I argue that Peirce himself formalizes this process—abduction and all—in the creation of his Existential Graphs. This final move forces me to depart temporarily from the chronological approach to Peirce's corpus. As Murray Murphey and others observe, Peirce does not begin his work on Existential Graphs in earnest until the late 1890s. This departure, however, will be brief. It helps us shed light on the relation between Peirce's logic and the unique metaphysical and ontological position he advances.

Earlier I claimed that abduction was the initiating moment of—or, more accurately, one recursive moment in—the process of inquiry. The discussion of abductive musement elaborates this suggestion, indicating that further investigation might be warranted. While certain models of abduction have proven inadequate to describe this process of insight-inference, another type has been developed recently that attempts to model abduction in terms of a process of inquiry instead of regarding abduction in terms of the hypotheses it produces.

This model stems from the work of the legal theorist John H. Wigmore in the 1940s and has been developed in detail very recently by David Shum and the computer scientist Douglas Walton. Wigmore's understanding of abduction, framed by his interest in constructing diagrams of legal argumentation, has been largely overlooked in recent accounts of Peircean logic. But it seems to be in line with the investigation of the imagination. Wigmore's work stands to place the study of abduction on a more fruitful course. Schum distinguishes his diagrammatic approach to abduction from that of computational models such as PEIRCE. "As Wigmore realized, the construction of an inference network is an exercise in *imaginative reasoning*" and is a matter of "how one might construct different plausible chains of reasoning from evidence in a particular case."[27]

Walton elaborates on Wigmore's understanding of abduction as a moment in the process of inquiry and clarifies Wigmore's attempt to represent this process through the construction of a diagram. Walton writes:

> The argument diagram begins with a particular case in which the argumentation is given as a product, so to speak. There is a given text of discourse in which some argument, or perhaps an explanation has been put forward. The task is to state the premises and conclusions in the chain of argumentation, filling in the inferential steps joining them, along with the missing statements needed to make sense of the argumentation, and build up the argument diagram. The problem, however, is that abduction is an *imaginative process related to discovery*. Thus to analyze cases of abductive reasoning, *we need to go beyond the notion of argument as a product and view the argument as a process*.[28]

I draw out several points in this description, as follows.

1. Wigmore's argumentation diagrams depict an agent's possible response to a particular legal case. More generally, the diagram symbolizes an interaction between an investigator, a set of confusing circumstances, and a field of possible explanatory assertions. To put the point somewhat

differently, the diagram originates in and responds to a uniquely problematic situation.

2. The solution to this problematic situation is not readily at hand or, more accurately, is not recognized as being readily at hand.

3. The initial explanation an investigator provides is a hypothesis that is not derived from any previous postulates. Rather it is merely suggested as a possible answer by the context of the case. That is also to say that this explanation is ampliative and contextual. This seems to coincide with Peirce's attempts to describe the character of abductive processes. In an early commentary, W. M. Brown describes the plausibility of an abductive suggestion: "Plausibility, unlike Truth, is relative to a particular stage of inquiry. If Peirce is right, the relevance of a hypothesis is determined in part by what sort of phenomena we are trying to account for, and its general acceptability is relative to the current state of our budget, time, money and energy."[29] To put Brown's point slightly differently, the acceptability of abductive guesswork rests not on given postulates or preestablished benchmarks but on the particular purposes of an investigator's query.

4. Once an initial hypothesis is formed, it provides direction to future inquiry, providing both the limiting and enabling conditions of future investigation. Choosing a hypothesis is, *for the time being*, an irrevocable fact that necessarily precludes other possible explanations. These other possibilities are, at least temporarily, overlooked as the given hypothesis is explored.

5. The process of investigation is open to the criticism of others. Indeed, just court proceedings are just only to the extent that the proceedings provide a rational justification of a particular hypothesis, a justification agreed upon by a chosen set of peers.

With these points to guide my investigation, I turn my attention to Peirce's development of Existential Graphs (hereafter EGs). First, I provide a brief description of Peirce's Alpha Graphs. I then begin to explain the way in which Peirce regards these graphs as a means of *demonstrating* query-driven processes. Here we find a way of formalizing different aspects of imaginative inquiry without reducing these aspects to a closed logical system.

Peirce's Alpha Graphs: Representations of Imagination and Inquiry

Peirce wrote his first EG paper in 1882 and continued this project until his death in 1914. He envisions the construction of these graphs as a way of analyzing and mapping certain types of inquiries. The first part of the EG system is termed the Alpha section. It is consistent with two-term Boolean algebra and can visually model standard propositional logic. The best way to understand the Alpha Graphs is by an examination of their syntax and semantics. By the syntax of graphs, I am pointing to the *conventions* of the Alpha section. By the semantics, I am pointing to the *meaning* and philosophic import of these conventions.

The syntax of Alpha can be described in the following way:

1. Sheet of Assertion (SA)—the blank page.
2. Assertions—single letters or phrases written anywhere on the page.
3. Object enclosed by a closed curve (a cut). A cut cannot be empty; they can be nested but will never intersect.
4. Any well-formed part of a graph is termed a subgraph.

The semantics or meaning of the graphs can be sketched out in the following manner:

1. The SA denotes the field of possible truth.
2. Assertions—letters, phrases, and subgraphs—can be either true or false. (This fact allows graphs to represent the binary character of Boolean algebra—Boolean complementation.)
3. To surround assertions with a cut is to negate logically that assertion—an empty cut denotes "false."
4. Note that assertions enclosed *by the same cut* are brought into a particular relation—they are logically conjoined.
5. The aforementioned rules and meanings create a system that can visibly demonstrate the following logical relations: assertion, negation ($\neg$), conjunction ($\wedge$), disjunction ($\vee$), and conditional statements ($\rightarrow$).

With these conventions and meanings in mind, we can begin to construct some very elementary Alpha Graphs and provide their logical equivalences, as shown in Figure 1.

6. These graphs can be constructed as shown in Figure 1.

7. Note that the logical relations established by the graphs are estab-
    lished by a process of negation. For example, the conditional "If p
    then q" is established by the graph that literally reads, "It is not the
    case (false) that P is true and Q is false" or "Not (P and (not Q))."

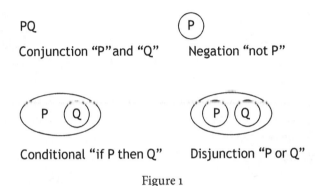

Figure 1

Kenneth Ketner's *Elements of Logic: An Introduction to Peirce's Existential Graphs* gives a succinct summary of the Alpha system. He
provides the general rules by which graphs can be transformed and
particular diagram transformations are proven to be logically valid.[30]
The first rule is the rule of the DOUBLE CUT. This rule states that a
double cut may be removed from under, or put under, any graph or
subgraph. Ketner rightly describes this rule in terms of the movement
of human inquiry over time. Underscoring the importance of logical
negation, he states that the rule of the double cut is best exemplified by
an inquirer who changes her mind over time. Initially, this inquirer
makes the assertion that "All plants need water to grow." She places a P
on the SA in order to reflect this belief. After a period of experimentation with cacti, she changes her mind and decides that "it is not the
case that all plants need water to grow." She places a cut around the
first assertion in order to negate this belief. Finally, after further study,
she reverses this latest denial and reasserts that "All plants really do
need water to grow." She can do this by placing another cut around the
first cut she made, hence making a double cut. Peirce observes that
there is a particular equivalence between the first assertion and the
double cut.

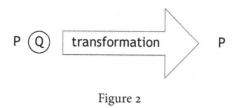

Figure 2

The second rule is the ERASURE. This rule holds that any graph or subgraph that is on an even level (2n) may be erased. As Ketner states, "in order to use this rule we need to agree upon the notion of a level. . . . First we must consider SA, the level on which we write all of our graphs, as level two, an even level." He suggests that we consider levels as "hockey pucks" that get stacked on the SA and on one another. This rule allows for the transformation of two graphs that represent the statements "assert P and deny Q" and "assert P." Graphically, they would be written as shown in Figure 2. We can check to see that this transformation is truth preserving by assuming that the transformation is not true and tracing the implication of the assumption until we reach a contradiction.

The third rule of DEITERATION states that within any continuous level, a graph or subgraph that appears more than once may have all except one instance removed. If a statement appears on higher and lower levels of a given graph, the statement on the higher levels can be removed (as long as no depressed sections between the two assertions are crossed). The first part of the rule holds because within any given level, any assertions that are repeated are equivalent. Let us consider the second part of the rule graphically. It states that the transformation shown in Figure 3 is truth preserving. Again, we can prove that this is truth preserving by assuming that the transformation is not true and tracing the implications of the assumption until we reach a contradiction in terms. One more comment is warranted in regard to deiteration. Consider the graph that represents "P and deny (P and R)." P is on both level 2 and level 3. There are no depressed levels between the two statements. Remember that a cross-section of the graphs has a certain "depth." If there were a depressed region between these statements, the transformation would not be truth preserving.

The fourth rule of the Alpha system, ITERATION, follows from the third. Iteration states that "within any continuous level, a graph or subgraph may be repeated; and a graph or subgraph appearing on a lower level may be repeated on a higher level, provided no depressed areas are crossed."

Peirce terms the fifth and final rule INSERTION. Insertion holds that any well-formed graph can be inserted into any part of odd level (2n+1) given that nothing moves or is covered. Insertion states that the transformation shown in Figure 4 is truth preserving.

Having briefly described the rules of the EG system, we can begin to employ the Alpha system to prove/test certain logical transformations. Consider the proof of *modus ponens* (mode of positing) through EG transformations (Figure 5).

It is important to note that graphs are transformed rather than rewritten. Transformation and proof construction are to be viewed in a sequence that Peirce believes provides a moving picture of human inquiry. This picture can be "played," "rewound," and "fast forwarded." It is a scene of activity. In his notebooks entitled "The Basis of Pragmatism," Peirce states that the definitions of his EGs will "be given in strictly pragmatic form, that is in the form of precepts of conduct, more definitely speaking, as permissions to do certain things under expressed general circumstances."[31] The dynamic character of the EG is examined in Figure 5.

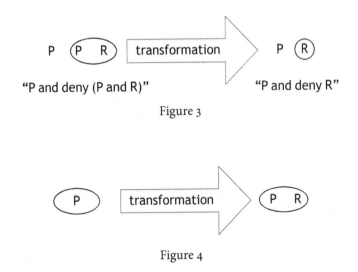

"P and deny (P and R)"   "P and deny R"

Figure 3

Figure 4

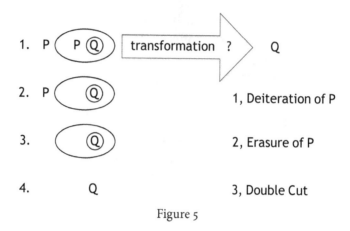

Figure 5

Graph Transformation as the Logic of Question and Answer

In 1907, Peirce wrote that "existential graphs constitute the most perfect system hitherto proposed for the analytic representation of ratiocinative thought."[32] Why is this the case? How does a series of sheets marked with letters and ovals represent the most "perfect system" for the analytic representation of thought? It may be possible to understand this bold claim and answer the aforementioned questions in reference to the Alpha system.

As opposed to the calculus, which allows us to move from a premise to a given conclusion in the fewest number of steps, Peirce's graphical logic, demonstrated in Alpha, allows us to examine and represent the *process* by which one eventually arrives at a given conclusion. What diagrammatic thought lacks in expediency it more than makes up for in integrity. Unlike the calculus, Peirce's graphs faithfully symbolize the *movement* of human inquiry, taking account of the imaginative elements that unexpectedly emerge in its continuous process. It is important to stress a point that is often overlooked in the study of Existential Graphs. Peirce does not intend his graphs to be regarded as static products but as a *mapping* of human processes. In the early 1900s, he remarked that graphical transformations should be scribed on separate sheets of paper so that they can be bound in a type of flipbook that could be used to scan the sheets quickly and in order, creating a "moving picture of the mind in reasoning."[33] Six years later, in a notebook on the "Prolegomena to an

Apology for Pragmatism," Peirce writes that "the extraordinary thing of Diagrams is that they show—as literally show as the Percept shows the perceptual judgement to be true—that a consequence does follow and, more marvelous yet, would follow." He explains that "it is not the static diagram that shows this, but the Diagram-icon having been constructed with Intention."[34] Pierce envisions EGs as a means to "show" a living continuity between the various stages of human inquiry. This flipbook provides a mapping of a particular inquiry and invites the participation of others. As a person flips through a series of Alpha sheets, she is drawn into the movement of inquiry, and, indeed, her participation motivates the dynamics of Peirce's graphs.

With these preparatory remarks in mind, it seems safe to draw out several key features of Peirce's Existential Graphs. As Ketner and others observe, EGs are primarily concerned with "assertions." "An assertion occurs when a person claims that a particular statement is true and indeed accepts responsibility for that statement."[35] This echoes Peirce's unpublished statement in his notebooks on pragmaticism: "To assert a proposition means to accept a responsibility for it so that if it turns out ill . . . which is called proving it to be false, he who asserted it regrets having done so."[36] This assertion emerges as a result of an agent's decision but also by virtue of the possibilities contained in a particular situation of inquiry. In terms of EGs, Peirce describes this field of possibility as the Sheet of Assertion (SA). The SA is often described as a type of empty set, as a blank slate. Ketner, I think rightly, points out the inaccuracy of this reading.

> It is important to learn that the [SA] isn't really empty. Every point on the sheet is understood by us to mean a kind of invisible phantom sentence always taken to be true: "You may assert something here." So [the] SA is not really empty; it is teeming with an infinite conjunction of an infinite number of that phantom true sentence.[37]

Assertions arise in the interaction between an inquirer's purposes and this field of possibility. Peirce rarely addresses the fact that the SA actively permits assertions, but this point will prove important in describing the reciprocal relationship that characterizes imaginative inquiry. It is also worth noting that the possibilities of the SA are inexhaustible and cannot be prefigured.

In our discussion of abduction, it is necessary to note that any assertion made in an EG is made as a tentative hypothesis and "has the status of a 'might be true.'" Any statement written on a particular sheet may be considered only provisionally true, as a statement to be tested and "tried out" during subsequent EG transformations. Similarly, negations, known as "cuts," remove particular statements from the field of assertion. By enclosing a statement P on the SA, I take responsibility for the hypothesis that P is really false. These abductions—both positive and negative—emerge in the continuous course of investigation over a given span of time. While a single sheet of EG cannot demonstrate this point, the "moving picture" of thought created by a series of EG sheets demonstrates that abductive statements are continuous with past cuts and assertions. While they are continuous with the past, they are not necessarily *determined* by the past relations of inquiry. The choice of the agent and the permission of the sheet are present in the development of any EG transformation. This interpretation begins to expose the imaginative underpinnings of the EG system. Peirce's graphs allow us to see, experience, and reconstruct the way in which the old and familiar are made new in experience.

It is important to note that cuts and assertions may be hypothetical, but they have real consequences on the Sheet of Assertion. In other words, the content of the assertions and negations is provisional, but the fact that these actions are taken is an irrevocable reality, a reality that guides the course of future investigation. This point is apparent in the case of EG cuts. Denials preclude particular future assertions and, in so doing, delimit the space of possible investigation. This is seen in both the constraints placed on insertion and the guidelines of iteration-deiteration. The harmony of the "moving picture" of ratiocination is established by virtue of the responsible choice of an agent whose freedom has been guided and enabled—one might be wrongly tempted to say curtailed—by past assertions/cuts.

It is easy to overlook the imaginative character of Peirce's Existential Graphs. Peirce, after all, intends that these diagrams be employed as logical proofs that secure or, more accurately, demonstrate definitive conclusions. The various ways by which these conclusions are represented, however, are by no means definitive. The transformations that occur *between* premises and conclusion are realized in experimental and creative processes. Any novice logician can attest to the fact that proofs—even the line proofs of first-order logic—involve a type of guesswork and musement. It involves witnessing the field of possibilities, allowing particular logical possibilities to show themselves as viable candidates for testing, and, finally, the willingness to test these candidate hypotheses. The many erasure marks of a first-year student's line proof attest to the fact that logical hypotheses arise as tentative guesses and that the ability to witness, recognize, and select possibilities is developed over time. These points are explicitly worked out in Peirce's discussion of EGs and diagrammatic thought more generally.

According to Kathleen Hull—and Peirce himself—inquiry, logical and otherwise, begins with experience. Peirce restates this at multiple points, most notably in 1903 by paraphrasing Aristotle: "Nothing emerges in meaningful conception that first does not emerge in perceptual judgment."[38] Carl Hausman echoes this point, writing that according to Peirce, "all the sciences of discovery, among which he includes mathematics, rest on observation (1.239–40)."[39] This form of observation is described as phenomenology later in 1903, when Peirce suggests that we are "to open our mental eyes and look well at the phenomenon and say what are the characteristics that are never wanting in it, whether that phenomenon be something that outward experience forces upon our attention, or whether it be the wildest of dreams."[40] It is this event-process that Peirce intends to capture in his EGs and in diagrammatic thought more generally. The aim of phenomenology is to muse on an array of phenomena and to witness the appearance of structures common to these phenomena. Peirce comes very close to anticipating the definition of phenomenology as the process of "returning to the things themselves." For Peirce, phenomenology is an active seeing or

> study which, supported by direct observations of phanerons and generalizing its observations, signalizes a very broad classes of phanerons; describes the features of each; shows that although they are inextricably mixed together that no one can be isolated, yet it is manifest that their characteristics are quite disparate; (and) then proves beyond question that a certain very short list comprises all of these broadest categories of phanerons that are.[41]

A few points stand to be drawn out of this passage. First, phenome-nology is a form of direct observation of the phanerons. Phanerons are the collective sum of all that is possible in experience. Here one must be careful not to suggest that direct observation can seize the phaneron in its totality, just as no cut or assertion can exhaust the possibilities of the Sheet of Assertion. It stands as the collective whole that could be present in the mind.[42] Second, this direct observation is at once a "generaliza-tion" that "signalizes" classes within the phenomenological field. It is through this observation-generalization that one arrives at the classes or categories of Peirce's system. It seems reasonable and indeed necessary to use this phenomenological characterization in any reading of Peirce's logical diagrams.

This relatively early insight into the phenomenological character of the categories appears again in Peirce's development of logical diagrams that simulate the phenomenological method. These diagrams provide a formal and indirect schematic of the classification of the phanerons. In the last of his Cambridge lectures, Peirce suggests that such graphs can indirectly stand for the true continua of quality that is experienced in any phenom-enological musing.[43] We can now return to and complete Peirce's com-ment: "Mathematical truth is derived from observation of creation of our own visual imagination . . . *which we may set down on paper in form of dia-grams.*"[44] While inquiry begins in musing about the multitude of actual and possible phenomena, it takes flight in the creative play of the imagina-tion that infers certain resemblances beneath the field of phenomenological possibility. This type of inference-insight is operative throughout any trans-formation of EG. Peirce begins to hit upon this peculiar type of inference in 1893, commenting that "as experience clusters certain ideas into sets, so does the mind too, by its occult nature, cluster certain ideas into sets."[45] While this "occult" operation of the mind may not be deductive, Peirce comments that this process of seeing resemblances ought to be considered as a form of logical inference and rests at the heart of diagrammatic experimentation. The musement of the imagination, in light of an array of phenomena, gives rise to a form of instinctive-intuitive reasoning that ex-poses the possible relations between the individuals of the multitude. More simply yet more impressively, this occult power has the intuitive ability to move from a given set of phenomena to a more holistic conception, a conception of which the given collection is but a part. Peirce repeatedly employs diagrams (of broken stars, of incomplete circles, of fragmented boxes, and most notably of EGs) to illustrate our power to imagine the possibility of a whole from a fragment. This movement from partiality to generality is demonstrated by EG proofs. Not incidentally, it is also this movement that characterizes the abductive leap that is often made from particular experiences to general explanatory hypotheses.

Mathematicians and logicians make the intuitive leap from several elements to the set of which the elements are members; that is, they move from concrete examples to general rules. Intuition allows both mathematicians and logicians to infer the possible reality of more complex sets from the existence of less complex collections. In terms of EGs, Peirce suggests that there are two distinct means of graphical transformation, which he describes as corollary and theorematic reasoning. It is the theorematic aspect of the transformations that reflects the ampliative character of intuition. Peirce writes that corollary reasoning proceeds from fixed premises and, in terms of a particular diagram or graph, makes explicit the relations between elements already present in the diagram. Theorematic transformations, on the other hand, obtain when an investigator *sees* and *infers* elements that are suggested by but are not present in a series of EG sheets. Peirce comments to this effect, stating that the "peculiarity of theorematic reasoning is that it considers something not implied at all in the conceptions so far gained, which neither the definition of the object of research nor anything yet known about could of themselves suggest, although they give room for it."[46] In his *New Elements of Mathematics*, Peirce underscores the distinction between theorematic and corollary reasoning as "a matter of extreme importance for the theory of cognition."[47] He states that any corollary would be a proposition deduced directly "from propositions already established without the use of any other construction than one necessarily suggested in apprehending the annunciation of the proposition."[48] By comparison, Peirce makes the imaginative and ampliative character of theorematic reasoning explicit.

Any *Theorem* (as I shall use the term) would be a proposition pronouncing, in effect, that were a general condition which it describes fulfilled, a certain result which it describes in a general way . . . will be impossible, *this proposition being capable of demonstration from propositions previously established, but not without imagining something more than what the condition supposes to exist.*[49]

This distinction has proven to be a matter of extreme importance for contemporary researchers in the field of artificial intelligence. While automated theorem models can perform corollary functions more efficiently than any human investigator, they are easily outmatched by average high school students when asked to perform theorematic tasks. The question of theorematic reasoning and the difficulty faced in modeling imaginative processes seems to throw one back on the particular embodiment and willing attentiveness that is unique to living organisms.

In terms of our current investigation, this nondeductive, theorematic form of reasoning could also be described as a type of Peircean abduction. More strongly put, we identify mathematical-logical intuition with the abductive creation of new hypotheses. Without delving once again into the details of abduction, it is sufficient to note that this inference takes the form of a hypothesis in reference to the possible relations that emerge in the multitude of experience.[50] Peirce envisions abduction as the way in which these hypotheses are formed. He repeatedly underscores the fact that only this novel guesswork can supply novel insight into the possibility of the continuous field of phenomena. Abductive insight recommends a course of *future* action and encourages an investigator to move forward on an experimentation of a given diagram. I provisionally suggest that this "moving forward" on a diagram, this "moving forward" in abductively prompted action, captures the essence of true continua. This is not to say that it captures the totality of true continua. Rather, intuitive hypotheses take account of true continuity in its principle of futurity. Mathematical intuition "moves forward," inferring the possibility of continuous set expansion into a presumably infinite future. The character of futurity and the principle of expansion-growth are constitutive of both true continua and diagrammatic thought.[51]

Diagrammatic signs, as one example, are triads or icons that exhibit "a similarity or analogy to the subject of discourse." This similarity or analogy is crucial since the diagram may be finite while its object is infinite.

When the diagram's object is infinite, it cannot be apprehended in its totality except by observing an analogous, finite structure upon which our intuition acts. Moreover, this resemblance or similarity is in terms of rational relations between the object's parts.[52] Thus, as Peirce states, "the pure Diagram is designed to represent and to render intelligible the Form of Relation merely."[53]

An EG diagram is a construction that is defined by one or more abstract precepts. It is on this basis that Peirce equates a drawn geometrical diagram with "an array of algebraical symbols." If the diagram is literally drawn or written, "one contemplates the Diagram, and one at once prescinds from the accidental characters that have no significance." Once laid out, the construction offers a mathematician, logician, or inquirer the opportunity to see relations "between the parts . . . or array that are not explicitly expressed by the abstract precept[s]."[54] For example, the diagram of the positive integers written one after another follows the abstract precept of "successor," and from that precept one can "skip" every successor and obtain the formal concept of the even numbers. These new relations are always obtained by rule-bound transformation, which is what Stiernfelt calls the "defining feature of the diagram . . . [which makes] it the base of [thought-experiment], ranging from routine everyday what-if to scientific invention."[55]

Even finite objects like triangles must have some generality. They must apply not only to the concrete diagram that has been written on a blackboard but also to all triangles generally. The diagram is a schema that is applicable to the infinite collection of empirical triangles. It is in this form of generalization that reason discloses the real possibility of true continuity. True continuity, according to Peirce, is "an indispensible element of reality, and that continuity is simply what generality becomes in the logic of relatives, and thus, like generality, and more than generality, is an affair of thought."[56] Henry Wang underscores Peirce's statement that "generality, therefore, is nothing but a rudimentary form of true continuity. Continuity is nothing but perfect generality of the law of relationship."[57] Wang's comment echoes Peirce's suspicion that true continuity cannot be understood in terms of a multitude of discrete individuals but instead as an array of qualities that demonstrates the principle of generality. These qualities are generalities. As Peirce states, "they are mere logical possibilities, and possibilities are general, and no multitude can ever exhaust the narrowest form of general."[58]

The intuitive reasoning that witnesses this generality-possibility, this "rudimentary form of true continuity," is not discrete in the sense of advancing from the abstract precepts that define its logical structure. Rather, the logician goes beyond the given precepts by developing hypotheses that expose the possible relations contained in the diagram. The mathematician does not prove this generality but rather experiences it in intuition. Hull and others suggest that these hypotheses ought to be considered "new intuitive idealizations," that is, as products of the creative imagination. It is in this sense that Beverly Kent suggests that "Peirce attributed creative thinking to the mental manipulations of diagrams."[59] By extension, this creative thinking is the heart and soul of EG transformations. It is worth noting that these mental manipulations have a meaning that goes beyond the revision of particular sign configurations on the basis of enumerated rules. These manipulations are not the products of some brain in a vat but are embodied and sensuous manipulations. Diagrammatic manipulations accompany an ongoing experience; they require an anticipatory involvement. They demand a particular bodily attentiveness.

Framed in this manner, diagrammatic thought reflects the imaginative legacy that Peirce inherits from Kant and Schiller. Diagrams provide Peirce yet another way of understanding the novelty and continuity that had long been associated with the aesthetic imagination. Peirce, however, draws this aesthetic and ampliative process into his logic, situating it at the heart of his formal system. EG transformations, I argue here, require a reflective process that arises between an investigator and the possibilities that a particular situation affords. The reader will notice that I have neither concentrated on the particular rules of Alpha, Beta, or Gamma graphs, nor have I paid attention to particular EG proofs. Instead, I have tried to situate EGs within the framework of Peirce's broader philosophy, exposing the way in which we might think of EGs as a formalization or representation of imaginative thought. Rather than focus unwarranted attention on particular EG transformations—a project exhaustively undertaken by Ketner and others—I have concentrated on the abductive/intuitive processes that emerge in transformations. These processes depend on a reciprocal transaction between an agent of inquiry and what Dewey would later call the "affordances" of an agent's context. The processes arise in a genuine transaction, in the nexus between the possibilities of the environment and the purposes of the individual inquirer—to extend the discussions of EGs—in the interaction between a logician and the sheet of possibility.

At this point, I note that there is no real distinction to be made between purposes and possibilities. Both actively present themselves, and both attentively await actualization. A good investigator—scientist, logician, or gardener—must carefully wait and witness the possibilities that show themselves in the play of the investigation. Peirce turns to this investigator's ability to wait, witness, and act over the course of human inquiry. Concentration on this ability enables him to ask once again how it might be possible for human beings to think imaginatively. As a result, like Kant, he begins to reconsider the natural basis of the imagination, the odd fact that imagination emerges in the embodied life of human beings who are constantly conversing with their natural surroundings. He begins to ask how spontaneity and irregularity might be possible in the ordering of natural processes. Having examined the epistemological and logical workings of abduction, he turns to the ontological and natural ground that provides the necessary basis for these workings. I now turn to consideration of this ground.

SIX

ONTOLOGY AND IMAGINATION:
PEIRCE ON NECESSITY AND AGENCY

Taking Chances: Tychism

During the 1880s, Peirce employed the triadic nature of thought to ground his budding cosmology. As Karl-Otto Apel suggests, it was during this time (particularly in 1885) that Peirce developed a "metaphysics of evolution."[1] Peirce's attempt to expose a continuous relation among the three aspects of human thought becomes a desire to unify three realms of being. He comes to reassert the necessary connection between epistemology and ontology. Just as Kant's discussion of imagination and reflective judgment in 1792 leads him to speculate on the topics of time and purposive nature, Peirce's examination of the triadic character of logic and inquiry—in particular the instinctual and inferential character of abduction—draws him, almost exactly one hundred years later, to a comprehensive rethinking of the order of the natural world. Logic must be fitted to the nature it seeks to describe. Nature must lend itself to inquiry.

This development in Peirce's thought is first seen in the "Doctrine of Necessity Examined" and "The Order of Nature." But it takes center stage in "A Guess at the Riddle" (1887), "The Law of Mind" (1892), and "Evolutionary Love" (1893). Most commentators take one of two approaches to these works: either they overlook the idealist overtones that define them, or they recognize the overtones and on these grounds dismiss the articles as flights of Peirce's metaphysical fancy. They emphasize Peirce's empirical and scientific interests at the expense of understanding the unique idealism that may, in the end, be compatible with these interests.

The analyses of these articles have been responsible for generating what is often called the "two Peirce hypothesis," stated plainly by Thomas Gould in 1950 when he suggests that Peirce proposes a hard-hitting empiricism that is inconsistent with the metaphysical speculation that emerges in his later works. I, however, resist this two-pronged analysis and instead attempt to show how Peirce's logical and epistemological accounts compliment, even if they are not wholly congruent with, his metaphysics.

In the 1880s, Peirce's emphasis on abduction and the spontaneous character of thought suddenly shifted to spontaneity as it shows itself in the natural world. Having encountered the work of Spencer, Darwin, and Lamarck in his early career, Pierce attempts to explain how the evolutionary history of the natural world might be carried forward in new and creative ways. This shift in Peirce's thinking results in his development of the doctrine of absolute chance, a concept he terms "tychism." As our discussion of diagrammatic thought and Peircean inquiry indicates, there must be a type of continuity between the natural sphere and the form of human inquiry that seeks to examine it. One important point of continuity, according to Peirce, is a type of common spontaneity. Peirce makes this point explicit when he suggests that his study of spontaneous events led him to develop the theory of continuity in his evolutionary cosmology. "I have begun by showing that *tychism* must give birth to an evolutionary cosmology, in which all the regularities of nature and of mind are regarded as products of growth."[2] If abduction resides in the middle ground between necessity and chaos, there must exist natural processes and actions that correspond to this unique and (I argue) imaginative spontaneity. Peirce begins to describe spontaneity's character and ubiquity in the "Doctrine of Necessity Examined."

> By . . . admitting pure spontaneity or life as a character of the universe, acting always and everywhere though restrained within narrow bounds by law, producing infinitesimal departures from law continually, and great ones with infinite infrequency, I account for all the variety and diversity of the universe, in the only sense in which the really *sui generis* and new can be said to be accounted for . . . variety can spring only from spontaneity.[3]

Again, it is important not to confuse spontaneity with randomness. The newness that Peirce highlights is "restrained within the narrow bounds of law" and is manifest in the nonlinear dynamics of complex systems.[4] Because of the connectivity that continuity implies, small adjustments in input can have dramatic effects on the development of a system.

While restrained by the bounds of law embodied in the stable structure of an organism, chance events proceed—literally "move forward"—as departures from this selfsame law. Peirce identifies a principle of irregularity that motivates growth and adaptation. In the wake of Darwin's discoveries, Peirce believed that the fact of irregularity and variation cannot continue to be overlooked. In an unpublished reflection on Darwin and Lamarck, Peirce writes that investigators who ignore the spontaneity of nature are "thinkers who overlook . . . overwhelmingly manifest, positive evidence." He asks, "What is the most obtrusive fact in nature? It is something which is overlooked because it is so plain. It is the variety and multiplicity of nature."[5] Peirce does not explain where this multiplicity comes from. But he does tell us that it "has not been produced by the operation of law." Furthermore, Peirce explains that to deny the variation of nature is to foreclose the possibility of nature's increasing complexity.

> To prescribe that under given circumstances a fixed result shall occur is to prescribe that the substantive manifoldness of nature shall never be increased. All this multiplicity, then, was either the work of a spontaneous will, or else it was without any definite cause, that is exists by its own spontaneity.[6]

As opposed to chance as pure contingency, this sort of irregularity is irreversible and hence directed. By identifying this general fact of irregularity, Peirce accounts for the variety and the diversity of the universe, but not in a way that can precisely predict the departures from the law. If spontaneity does not connote randomness and, at the same time, cannot be accounted for by mechanical necessity or linear computation, Peirce must identify "another form of causation, such as seems to be operative in the mind in the formation of associations."[7]

Peirce regards this form of natural causation as a particular type of agency. Exploring the loci of this agency is one of the objectives of the coming chapter, in which we turn to the empirical sciences and their study of human cognition. My suggestion concerning agency is not necessarily original. It is made by Hausman and indeed it is made by Peirce himself in the "Doctrine of Necessity Examined," where he writes that "in nature [there] is probably 'some agency' by which complexity and diversity is increased."[8] Here, Peirce does not succumb to what commentators occasionally and inaccurately describe as an Emerson-like theism. Rather, he succumbs to an Emerson-like disposition that takes seriously the individual purposes of the natural world. But the *agency* of nature? The mere suggestion prompts a type of incredulity that reflects the ambiguously sanctified character of the concept. Agency has, for better *and for worse*, been reserved to describe the freedom of the modern human subject. For this reason, our suggestion that the chance events that emerge in the dynamic unfolding of nature ought to be regarded as the actions of a type of agency warrants a more detailed treatment. Stuart Kauffman suggests that the study of autonomous agents will form the foundation of general biology in the coming decades. He defines this agency broadly:

> Consider a bacterium swimming upstream in a glucose gradient, its flagellar motor rotating. If we naively ask, "What is it doing?" we unhesitatingly answer something like, "It is going to get dinner." That is, without attributing consciousness or conscious purpose, we view bacterium as acting on its own behalf in an environment. The bacterium is swimming upstream in order to obtain the glucose it needs. Presumably we have in mind something like the Darwinian criteria to unpack the phrase, "on its own behalf."[9]

Kauffman explains that all "free-living cells and organisms" are most certainly autonomous agents to the extent that each cell is a physical system that can act "on its own behalf" in its environmental situation. With this point in mind, let us return to the discussion of Peirce and tychism. In Peirce's earlier writings, he suggested that spontaneity *without cause* could account for the diversity and growth of the organic world. By the 1880s, however, he had begun to revise his position on tychism—his doctrine of chance—concentrating on the spontaneity of purposive action and the possibility that chance events (like novel hypotheses) could be accounted for by way of "psychical actions." In 1906, Peirce explained this position in the unpublished "Prolegomena to an Apology for Pragmatism":

> I intend, as soon as I can command the requisite leisure from pot-boiling, to revise my tychistic hypothesis. I still believe that the universe is constantly receiving excessively minute accessions of variety; but instead of supposing, as I formerly did, that these are causeless (chances), I think there is sufficient ground for supposing that they are due to psychical action upon matter . . . at present, the psychical researchers have certainly cast serious doubt on our old materialist theory without instituting any progressive method of research into the problem. In this situation, a happy working hypothesis might prove of the utmost service. It would be a pity that the human race should go down to its grave, to which it is visible drawing near, without [addressing] its principal problem.[10]

Peirce suggests that the question of chance events ought to be addressed by way of a conception of "psychical action," that is, by way of a conception of agency. Hausman is quite good on this point when he writes that Peirce's understanding of psychical action ought to be considered the "originative condition that is in some sense responsible for its own action."[11] Hausman's comment is significant insofar as it begins to frame Peirce's understanding of "some agency" in the context that is often used to describe a purposive action or choice. This said, the understanding advances neither a reductive materialism nor an uncritical panpsychism. Rather it stands as an invitation to think through what it might mean to regard nature *as if* it were purposive. In effect, Peirce invites a reader to join him on a very lengthy—perhaps indefinite—investigation to examine the loci of agency and its unique character. To return to his comment in his 1906 "Apology," the task of understanding the *nature* of agency and "psychical actions" should be regarded as the "principal problem" of the human race.

Taking Guesses: Peirce's Imaginative Ontology

In "A Guess at the Riddle," Peirce applies his understanding of logic and epistemology to the construction of a philosophic architectonic. Aptly titled, this piece stands as Peirce's first look at speculative philosophy. It is a guess at explaining the way in which triadic ordering—originally conceived in terms of formal logic—manifests itself in the fields of metaphysics, psychology, physiology, biology, and physics. His division of these fields is maintained and extended in the final section of this project, which aims to investigate the structure and movement of the imagination from the perspective of the empirical sciences.

Before delving into Peirce's direct arguments, it is important to consider an indirect argument that commentators neglect in their analysis of the "Guess" and later works. This indirect argument hinges on the character of guesswork itself. By it, Peirce attempts to expose a continuity between the thoughtful lives of human beings and the natural world in all of its aspects. This is why he undertakes such an ambitious study of the natural and physical sciences. This is why he hypothesizes— "guesses"—that the organization of natural being must exhibit the same structure, movement, and spontaneity that is exhibited in the mind. This is why he "guesses" that he can begin to show a continuity between matter and mind. It is, however, not the *demonstration* of this point but rather the *guess itself* that provides the "proof" of Peirce's position on continuity.

"A Guess at the Riddle" is a moment of abduction in its own right, a moment of fortuitous hypothesis generation. As a hypothesis, it is an initial attempt to transform the indeterminacy of a situation into a more harmonious state of affairs. It is a new guess at resolving the riddle of life, not in the abstract but immediately, in the life of embodied human beings. The operative verb is "re-solve," since Peirce believes that we are repeatedly called by novel events to make novel hypotheses. Our ability to do so, he believes, demonstrates something about the relation between ourselves and the world at large, namely, that it is a continuous relation. Our hypotheses are fitted to the circumstances that we encounter not by some random coincidence but rather by a particular logic that prefigures our experience and our guesswork. Peirce emphasizes the ubiquity and self-reinstating character of hypotheses: if one attempts to refute the principle of abduction, one is forced to make a conjecture or good guess in reference to the refutation of the principle. We, as embodied beings encountering novel and unexpected situations, cannot avoid the novel guesswork that Peirce describes. Likewise, we cannot avoid the logic or order that prefigures these hypotheses.

A possible objection may be made to this final point, that hypothesis formation rests upon a prefiguring continuity between a thoughtful human being and natural being on the whole. After all, is the beginning of inquiry not marked by a type of discontinuity, namely, a problem to be solved? At the risk of repetition, I consider the argument from another vantage point. Let us ask ourselves what is *necessary* in the formation of a hypothesis. What is necessary in making a guess at a riddle? Peirce believes that when one makes a guess, there is, implicit in this hypothesis, a commitment to a particular course of action. At the very minimum, there is a commitment to the guess itself. Remember that abduction is the only pen that can write on our "docket of new cases to be tried." It is the necessary condition of purposive action. This commitment to action is prefigured by a disposition of hope, expectancy, and trust. It is the nascent emergence of a disposition that will later be described as the disposition of *agape*, or generative love. The relation between abduction and *agape* will be drawn out in a subsequent section of this chapter. At the moment, it is important to concentrate a bit more attention on the relation between hypotheses and the proof of continuity. If the guesswork of abduction depends on disposition of trust, Peirce suggests that this trust rejects the ontological view that the natural world is an unaccommodating, alien, discontinuous realm. Peirce believes that the generation of fortuitous hypotheses indicates otherwise, not only a posteriori, as demonstrated in the empirical tests it initiates, but also a priori, as the very precondition of its occurrence. It is in these two distinct voices that "A Guess at the Riddle" is expressed.

Note that the aforementioned proof is unique in its open-ended and embodied character. One "proves" the fact of continuity in and by the process of abduction, in a series of guesses that motivate the course of our daily affairs. Admittedly, the hypothetical character of our actions is occasionally, if not predominately, obscured by the habits of our lives, habits that often lead us astray and set us adrift in the turbulence of experience. This said, Peirce believes that in meaningful moments of guesswork we are opened to a sympathy and cooperation that exist between our selves and the "external" world. This belief is repeatedly expressed, most notably perhaps in Peirce's "Neglected Argument for the Reality of God." This neglected argument, an argument that rests on the character and preconditions of abduction, is resurrected in James's "Will to Believe" and in Dewey's *A Common Faith*. Both of these writers stress the continuous relation between mind and matter that abductive processes seem to expose. Dewey's work, often characterized as "theistic naturalism," stands as a particularly interesting appropriation of Peirce's insight into the relation between abductive inquiry and the experience of continuity.

Dewey couches this relation in terms of religious attitude. "A religious attitude, however, needs the sense of a connection of man, in the way of both dependence and support, with the enveloping world that the imagination feels is the universe."[12] The newness and harmony that the abductive imagination generates is the most immediate proof for the continuity between a thoughtful organism and its environs. It is the imagination that establishes this continuity. Dewey rephrases this point somewhat differently in another section of *A Common Faith*.

> Hence the idea of a thoroughgoing and deep-seated harmony of the self with the Universe (as a name for the totality of conditions with which the self is connected) operates only through the *imagination*— which is one reason why this composing the self is not voluntary in the sense of an act of special volition or resolution.[13]

For Peirce and Dewey, the harmony that well-suited hypotheses seek to establish is not produced by some autonomous action of an isolated individual. Rather it is realized in the nexus of an organism and a broader environmental situation. It is realized—like any hypothesis—by the light of nature (*lume naturale*) and in a space that has been described as uniquely imaginative.

Now we turn our attention to the demonstration of mind-matter continuity, to the a posteriori aspect of the argument that Peirce develops in the "Guess." Peirce's treatment of the fields of natural and physical sciences in the "Guess" is cursory. Certain comments there, however, need to be unpacked in order for us to address "The Law of Mind" and "Evolutionary Love" without dismissing them as flights of idealism. The conflation of logic and other fields such as psychology is deeply disturbing to some commentators. K. T. Fann, for instance, comments that "in order to have a clear understanding of Peirce's theory of abduction," one must keep in mind the distinction between its logical and psychological aspects.[14] Fann's stance, however, betrays Peirce's attempt to develop a unified theory of thinking and biology. Fann seems to overlook the fact that in the "Guess," Peirce provides a possible account of the way in which formal logic grew out of psychological processes and, ultimately, from the ground of natural being. Vincent Potter puts this more accurately when he argues that Peirce attempts to show how the logical "law" of thirdness is "a living power" that guides and is guided by organic nature.[15] In this light, some conflation in Peirce among logic, psychology, and the natural sciences should, on Peirce's account, not only be tolerated but encouraged.

In the "Guess," Peirce once again describes the triad of psychology as the regions of immediate feeling, knowing ("polar sense"), and "synthetic consciousness." Peirce extends this observation, identifying the necessary continuity between triadic cognition and the structures of the human body.

> Granted that there are three fundamentally different kinds of consciousness, it follows as a matter of course there must be something threefold in the physiology of the nervous system to account for them. No materialism is implied in this, further than that intimate dependence of the action of the mind upon the body.[16]

Here it is extremely important to remember what is afoot in Peirce's investigation of physiology. He has already identified the triadic character of cognition. I have made the case that this model of human cognition is defined by the creative and spontaneous imagination. By the end of the 1880s, Peirce is searching for the source or lineage of the imagination, the corporeal basis of creative thought. To this end, he examines the workings of neural networks and multicellular organisms, highlighting the way in which impulse, the transfer of impulse, and adaptation define organic life. Peirce develops a description of protoplasm, for example, that defines it by its odd plasticity and novel responsiveness to underline a type of continuity between human cognition and other forms of biological organization. The continuity that he hopes to expose is not mere similarity. It is rather an explanation of how "higher" forms of cognition can arise from the physical stuff of the world. In other words, Peirce explains how complexity can arise from simple systems. He suggests that this very process of arising must have an ampliative or abductive character.

Peirce is after a type of evolutionary explanation.[17] One of the most striking characteristics of evolutionary theory for him is its attempt to explain how organic novelty is conditioned and enabled by a continuous history. This fact is reflected in the section "The Triad in Biological Development," in which Peirce touches upon population genetics and speciation. He explains individual variation, heredity, and natural selection in terms of his three logical/epistemological categories. In describing an evolving population, he outlines a system that appears to sense, respond, and adapt in an organized and seemingly purposive manner. He portrays a population, a community of organisms, as a single individual that seems to live, and move, and even think in a logic of ordered relations. At the same time, he tries, not always successfully, to explain how the evolution of a population always involves the variation and adaptation of its individual members over time. It is worth noting that here Peirce takes on the issue of the "one-and-many," an issue that has befuddled thinkers in the history of philosophy and that would prove formative in the development of James's work in the early years of the twentieth century.

In the "Guess," Peirce identifies a series of nested systems unfolding together over time. As a reader works through the article, the complexity of this nesting becomes apparent. Even a provisional description of natural being requires multiple levels of analysis. After some comments on psychology, physiology, and biology, Peirce extends his discussion to chemistry and physics, two levels of analysis that shed light on the first three. One might read the latter half of my present project as an extension and deepening of Peirce's "Guess." Beginning with the philosophic concept of the imagination, it proceeds with increasing empirical rigor in order to expose the embodied *principle* of this cognitive process.[18] As Peirce demonstrates, such a project leads *naturally* to a study of modern physiology, biology, and physics.

Even in his most heady cosmological speculations, Peirce *never* abandons his commitment to empirical research. Indeed, his empiricism exposes the root of his cosmology. His empiricism reveals a set of triadic relations that seems to unify the realm of thinking and the categories of physical being. This point is obvious in his work in 1892 and 1893. During these years he begins "The Law of Mind" and "Evolutionary Love," his "idealistic" works. But it is obvious that he is also committed to exploring the possibility of this unity between material and mental realms by way of empirical science.

From Guesses to a Law

Peirce opens "The Law of Mind" by admitting that he may be, against his better judgment, developing a cosmology "in which all the regularities of nature and of mind are regarded as products of growth and a Schelling-fashioned idealism which holds matter to be mere specialized and partially deadened mind."[19] He admits to contracting the transcendental disease that circulated in Concord in the mid-1800s, commenting that "after long incubation, it has come to the surface, modified by mathematical conceptions and by training in physical investigations."[20] We should remember that the title "The Law of Mind" is the same as the title of one of Emerson' University Lectures at Harvard in 1869–1870. These were the same lectures that featured a young C. S. Peirce, who spoke on the British logicians. It was only later in Peirce's career that he came around, once again, to the suggestion put forth by a very aged Emerson in this lecture. Many commentators regard Peirce's commitment to idealism as a type of aberration that threatens to distract future scholars from the serious work of Peirce's philosophy of science. I, however, view Peirce's commitment to figures such as Schelling, Schiller, and Emerson as the foundation of his philosophy of science. Indeed, the position that Peirce repeatedly takes that "matter is mind which has come under the almost complete domination of habit"[21] is the conclusion of his detailed investigation of physics and the other natural sciences. In an unpublished passage that resonates very closely with "The Law of Mind," Peirce describes his findings in the natural sciences:

> Thus these two elements at least exist in nature, Spontaneity and Law. Now, to ask that Spontaneity be explained is illogical, and indeed absurd. But that Law be explained is a natural and proper demand of the Spirit of Inquiry. And it cannot be pronounced inexplicable and ultimate without blocking the path of inquiry. But to explain a thing is to show how it might have been the result of something else. Law, then, ought to be explained as a result of Spontaneity. Now the only way to do this is to show in some way that law may have been a product of growth, of evolution. Then we must make some principle of growth more fundamental than any mechanical law ... the suggestion to which this leads is obvious. It is that matter is mind which has come under the almost complete domination of habit.[22]

True, this passage reflects the German transcendentalism of Peirce's educational upbringing. But it also clearly articulates Peirce's notion of synechism, the law of mind that states that ideas spread as a continuous unity and, in so doing, acquire generality. It is also the law that stems from and accommodates spontaneous growth and the evolutionary movement of being. Synechism grows out of tychism in a logical and evolutionary sense, describing the fruitful middle ground between *absolute* chance (*tuché*) and *absolute* necessity. Indeed, it accommodates both chance and a type of determinability. It describes the region in which imagination, hitherto described, comes into play (*Spiel*). The mediating play of the imagination is, in a very important sense, free. At the same instant, however, it is bound to a particular experiential context and to its particular history. At important moments, the freedom of the imagination can give rise to regularity. Imagination is a freedom-within-constraint. For Peirce, synechism describes the imaginative mind. But it also delimits the region from which living being itself arises. There is an important elision between mind and natural being that is repeatedly underscored.

Peirce parts company from most idealists in the development of this elision. At many points, he tries not to assert but to *describe* the character of organic continuity. In these considerations Peirce repeatedly employs his background in empirical science and mathematics. In "The Law of Mind," he tries to flesh out synechism, the underlying principle of mind and being, through a detailed examination of mathematical continuity. It is necessary to take careful account of Peirce's analysis. His contributions in the area of mathematics and science continue to ground a variety of contemporary cognitive models and help form the basis of complexity theory, which has recently been appropriated by the fields of biology and physics. He may have been on the trail of a "Schelling-fashioned idealism." But he has, on the way, provided analytic and empirical insights that remain surprisingly useful. On the way to making an astonishing number of empirical insights, Peirce stumbles on unified metaphysics and ontology.

Before turning to an investigation of the imagination, carried out by the contemporary empirical sciences, it is necessary to linger just a bit longer on the bold and far-reaching conclusions Peirce draws from his empirical studies. If we are to venture into the fields of cognitive psychology, biology, and physics, it seems appropriately Peircean to recognize the possibility that these studies may lead us to metaphysical speculations, ontological suggestions, and ethical beliefs. In 1893, after decades of intense scientific experimentation, Peirce sketches out his conception of "Evolutionary Love." This article stands as one of Peirce's most poignant attempts to address the metaphysical and ethical implications of his philosophy of mind. A reader familiar with Peirce's work from the 1890s, however, will hardly be shocked that he arrives at this point.

In "Evolutionary Love," Peirce concludes that synechism, the law of mind that states that individual ideas spread continuously over time, implies a principle of "agapasm," or the principle of generative love. As stated earlier, the term synechism explains the mind's continuity with the world at large (natural and societal) and accommodates both the freedom and the limiting conditions through which the imaginative mind operates. In earlier renderings of synechism, Peirce has trouble maintaining these two seemingly contrary tendencies in balance and looks to the principle of *agape* to demonstrate their compatibility. *Agape*—like hypothesis formation, abduction, and synechism—mediates between tychism (chance) and anancasm (necessity).[23] On this point, Peirce writes, "the movement of love is circular, at once projecting creations into independency and drawing them into harmony."[24] It is in this sense that *agape* has long been associated with a type of purposive growth that is neither random nor mechanistic. At the risk of repetition, this unique disposition is reflected in the description of the imagination provided in earlier sections. Indeed, *agape* and imagination go hand in hand.

As Anderson and Ventimiglia suggest, imaginative and creative thinking depends on conditions of agapastic love in order to flourish. One might notice that this is the same point that was made earlier in our discussion of the prefiguring continuity that is exposed in the generation of imaginative hypotheses. From Anderson and Ventimiglia's perspective, one might re-read Gottfried Jacquin's statement to Mozart that "neither a lofty intelligence nor imagination go to the making of genius. Love, Love, Love. That is the soul of genius."[25] The Peircean gloss holds that it *is* imagination that makes genius, but only to the extent that imagination bespeaks a type of love (*agape*). *Agape* is the reciprocal relation—between self and world, between the human and the natural—by which genius is made possible. For the purposes of the empirical studies addressed in the final sections of this project, Peirce's description of agapasm provides us a way of thinking through the reciprocal relation between emergence and aggregation that defines human thinking and organic life more generally.

Evolutionary Love: Abduction and Agape

Many commentators dismiss "Evolutionary Love" as a moment of idealistic weakness in Peirce's otherwise strong analytic corpus.[26] They point to Peirce's appeal to the conception of creative love as a unifying principle of reality as a complete departure from his detailed account of epistemology. I take issue with this interpretation and believe that a reexamination of his ostensibly metaphysical essay belies Peirce's empirical sensibilities. These sensibilities, while remaining staunchly committed to experimental methods, part company with the standard understanding of materialism that often employs this method as its handmaid.

In addition to the *Spiel* of musement, Peirce inherits from Schiller's poetry the concept of *agape*. Indeed, for Schiller, the creative act of *Spiel* must be accompanied and conditioned by a certain type of love. It seems likely that Peirce was not only exposed to the *Aesthetische Brief* but also the *Philosophische Brief*, a lengthy work that the young Schiller developed as a response to the materialism that defined the philosophical atmosphere of his day. At the heart of this work lies the *Theosophy*, and at the heart of the *Theosophy* lie the sections entitled "*Lieben*" and "*Gott*." In reference to the topics of love and God, Schiller explains that human creativity—most notably reflected in artistic genius—opens individuals to the generative force that animates nature. This is not a mechanical or deterministic force. Rather it is a unique form of love. We should not get sidetracked by Schiller's appeal to "God" in his discussion. Throughout the *Theosophy*, especially in "*Gott*," it becomes obvious that Schiller's God is an archetypal figure, an embodiment of a creative principle. This said, it remains curious, if not somewhat disturbing, that Schiller's God is an archetype with a human correlate. Has Schiller projected human creativity to the realm of the superhuman and elevated genius to the sphere of the divine? While this question ought to give us pause, it should not bar the way to investigating the creative principle Schiller describes. Schiller describes this principle as a reciprocal relationship between two agents, an existential relationship that cannot be reduced to the action of either party in isolation. The reciprocity involved in creativity is brought to the fore in the *Philosophische Briefe*. It is also evident in a letter that Schiller writes to Reinwald on April 14, 1783, in the midst of writing *Don Carlos*.

> Just as no perfection can exist singularly but rather shows itself in a *general purposive relation*, no thinking soul can satisfy itself in isolation and withdrawn in its own existence. . . . There is an inner disposition to be enveloped simultaneously by the other and to consume desirously the other that is love.[27]

Schiller's description of the creative attitude, one that performs a sacrifice in the very act of asserting itself, is the attitude of agapastic love. It sheds light on what Peirce might have meant when he claims that love encourages individuation while, at the same time, drawing things into

harmony. As many scholars observe, Schiller attempts to mediate between one form of Platonic love and the other—between the fellow feeling of agape and the sensuousness of Eros. This is important insofar as
it suggests that the emergence of novelty is not generated by an agent in
erotic isolation but arises in a feeling, transactional process that obtains
between agents.

Carl Hausman argues that the typical Christian rendering is not
necessary to understand Peirce's use of *agape*. I, however, think that its
Christian trappings at the very least give some direction to our study of
evolutionary love. Ventimiglia and Anderson observe that *agape* is open
to many interpretations because of its long and rather uncertain history
in the Christian tradition. This said, they suggest, I think rightly, that
the principle is "generally summarized as a selfless or nonacquisitive
love that seeks to nourish the growth and foster the welfare of its
object."[28]

While it is unwise to ally Schiller or Peirce too closely with a Christian ethic, it is worth briefly investigating their conceptions of *agape*
along scriptural lines. The prologue of the Gospel of John reflects Peirce's
philosophy of mind and leads naturally to a development of *agape*. John
opens with a description of the co-emergence of logic, in the form of the
Word (*logos*), and being. "In the beginning was the Word, and the Word
was with God and the Word was God. He was with God in the beginning. Through him all things were made . . . The Word became flesh and
made his dwelling among us."[29] Here, it seems appropriate to describe
the Word as the *logos*, the order and relation of things. If human logic
has the ability to apprehend the ordering of reality, its ability rests on the
logos that girds both the imaginative mind of human being and the purposiveness of being in its most general form.

It is in this same sense that Peirce remarks, "all matter is really mind."[30]
He holds that this logical continuity is felt as a particular "sympathy"
between human beings and between these beings and their lifeworld.
He elaborates on this point:

> The agapastic development of thought should, if it exists, be distin-
> guished by its purposive character, this purpose being the development
> of an idea. We should have direct agapic or sympathetic comprehen-
> sion and recognition of it, by virtue of the continuity of thought.[31]

Identifying the purposive nature of things is not merely a metaphysical flight of fancy. It is a necessary extension of Peirce's examination of the nature of embodied human reasoning.

That logical and scientific investigations draw Peirce to a particular metaphysical and ontological position becomes apparent once again in a series of logic lectures he gives in Cambridge in 1898. Speculations in reference to the purposiveness of nature emerge from his sketches of triadic logic. In the opening lectures, entitled "Philosophy and the Conduct of Life" and "Types of Reasoning," Peirce walks his audience through an abridged description of the types of human reason. In the process, he once again outlines the distinction between deduction, induction, and abduction (retroduction). Peirce is emphatic that the triadic logic he develops is no disembodied or highly esoteric form of reason. As the title of the first lecture indicates, his logic of relatives seeks to describe and negotiate the conduct of life. Like the process of imagination described earlier, his logic interprets, affects, and is affected by the world at large. This reciprocity between nature and the imaginative mind leads him to draw a conclusion that will be pursued in the final section of this book. Peirce concludes by reframing the words of the biblical Paul by converting a form of Christian theism into a panlogism of a very unique stripe:

> But if there is any reality, then, so far as there is any reality, what reality consists in is this: that there is in the being of things something which corresponds to the process of reasoning, that the world *lives,* and **moves**, and **HAS ITS BEING** in a logic of events.[32]

For Peirce, there is no divide between epistemological ordering and the ordering of the natural world. Both assume the same triadic structuring. Rather than repeat the mistakes of modern philosophy—the false dichotomy between mind and body and the mistaken distinction between form and matter—Peirce seeks to revive and clarify a Parmenidean fragment: "For the same thing is for thinking as for being."[33] It is the mediating process of the imagination, as first rendered by Kant and appropriated by Peirce in his philosophy of mind, that holds the key to this fragment of ancient thought. It is in the light of this comment that the title of Peirce's lectures, *The Reason and Logic of Things,* is seen in its full significance. In our return to the things themselves, we are awakened to the "reason and logic" that prefigure this return. Continuity and the possibility of order are found in, and created by, the world—in and by the relations of things—of which the human imagination (*ingenium*) is but another instantiation. This observation in reference to continuity grounds Peirce's later epistemology and semiotic. But, perhaps more importantly, it necessitates a particular system of ontology, metaphysics, and ethics.

Peirce, like Kant, seems to understand that his discussion of a continuity between mind and matter implies a radical revision of the natural sciences. By the same token, he understands that many scientists will be unwilling to accept that "things" have a particular reasoning and logic that is commensurate with what has long been considered the unique character of the imaginative human mind. Peirce concludes his lecture by reflecting on this understanding:

> I have not succeeded in persuading my contemporaries to believe that nature also makes inductions and retroductions. They seem to think that her mind is in the infantile stage of the Aristotelian and Stoic philosophers. I point out that Evolution wherever it takes place is one vast succession of generalizations, by which matter is becoming subjected to higher and higher laws [aggregation]; and I point to the infinite variety of nature as testifying to her originality or power of retroduction. But so far the ideas are too ingrained. Very few accept my message.[34]

The "message" that the purposive character of human thought arises from the purposiveness of nature, that this purposiveness can be described as a process of triadic mediation over time, is still extremely difficult for many to accept. Despite this, the foregoing epistemological description, focusing on the process of the imagination, originating in cryptic statements of the ancients, revived in the historical context of German idealism, and resurrected in the advent of American pragmatism, forces one to reconsider this suggestion. The final chapters of this book seek to reexamine this "message" from the perspective of the empirical sciences. This reexamination will be structured on a model that Peirce suggests in "A Guess at the Riddle."

For this message to be convincing, it will be necessary to describe the particular way in which the imaginative mind is embodied. Such a description is rooted in cognitive psychology and modern neuroscience. Both disciplines have made a sustained effort to highlight the way in which conceptual organization emerges from the physiology of the human organism. These disciplines lead us to an even more perplexing question that must begin to be answered if Peirce's message concerning purposive nature is to be taken seriously. How did the physical stuff of the world give rise to an ordered physiology that, in turn, over time, gave rise to human imagination?[35] This question brings the true scope of this project into focus. Thinking through the imagination implies the humbling project of thinking through the evolution of life itself.

SEVEN

THE EVOLUTION OF THE IMAGINATION

Remember the Sphinx: A Prelude to the Empirical Sciences

As we turn our attention to the explanations of the imagination furnished by the contemporary empirical sciences, it is wise to remember the Sphinx.[1] When Peirce wrote "A Guess at the Riddle," he asked that a small vignette of the Sphinx be placed under the title. But why? I suspect Emerson (the author of "The Sphinx" in 1841) and Thoreau (a commentator on this favorite poem of Emerson) had a pretty good idea. Once again, "A Guess at the Riddle" was Peirce's rough outline of the way that triadic relations obtained first in metaphysics and psychology and then in the natural world (physiology, biology, and physics). In traversing these fields, Peirce was intent on showing how chance, law, and habit formation were operative in the processes of mind and matter. This essay amounts to his response to the question of the Sphinx, to that riddle that continued to nag Peirce about the relationship between human creativity and the creative forces of nature. In short, it was Peirce's shot at answering the riddle of existence. No small task.

A quick lesson in mythology: the Sphinx—half woman, half lion—guarded the gates of ancient Thebes. It would ask travelers to answer a question, and when they were unable to do so, the monster would kill them. One day, a powerful man approached the Sphinx and was greeted with the following question: "What travels on four legs in the morning, two at noon, and three in the evening?" He responded: "Man, who crawls as a child, walks as an adult, and uses a cane in old age." Upon hearing this answer, the Sphinx threw herself into the ocean, freeing Thebes of her harsh rule. This man was Oedipus, who, thanks to the slaying of the Sphinx, was brought into Thebes as a hero and soon to be king. We all know that things ultimately do not go well for Oedipus. Indeed, his guess at the riddle of the Sphinx eventually reveals itself as partial, or incomplete, or inadequate, or disastrous. Even in providing a seemingly satisfying answer to the Sphinx, Oedipus commits unknowing and ultimately fatal mistakes. This is what we human beings do—and do with stunning regularity. The Sphinx was supposed to stand guard over Peirce's "A Guess at the Riddle" as a reminder of how mistaken seemingly good answers can be. It serves as a symbol of caution that might serve us well as we attempt to provide our own answers via the empirical sciences.

The previous chapter concluded with a question that motivates the remainder of this project: "How did the physical stuff of the world give rise to an ordered physiology that, in turn, over time, gave rise to the human imagination?" Working through the question of the imagination requires a particular humility, a particular type of doubt. It is the type of doubt that only the Sphinx can inspire. If, as Peirce and Kant seem to suggest, the processes of the imagination are at once continuous with the logic and structuring of nature, this humility ought to be readily apparent. The conclusions that we come to will not only bear on the field of aesthetics but on epistemology and the natural sciences on the whole. Indeed, if we proceed with utter confidence in our portrayal of imagination, it seems certain that it is not the imagination that will be described but some poor model that fails to acknowledge its scope and complexity. And we don't want to end up like Oedipus, do we?

To what extent is the imagination continuous with the natural world? Linguists, cognitive neuroscientists, and biologists such as George Lakoff,

Mark Johnson, Gerald Edelman, Donald Tucker, Antonio Damasio, and Vittorio Gallese, as well as complexity theorists and physicists such as Stuart Kauffman and John Holland, have begun to explore this continuity. In many cases, their work sheds new light on the imagination and provides new richness to the accounts that have been sketched out by Kant and Peirce.

The prior chapters began to describe the movement of the imagination: its receptivity, its novel activity, its ability to project into the future while remaining oddly rooted in its past. They have also described the movement of the imagination insofar as they tell a story about the development of the concept of imagination in the history of Western thought. This is a story that begins in the realm of aesthetics *qua* aesthetics, in the description of the imagination as a primarily artistic faculty. Relying heavily on the work of Kant and Peirce, I have attempted to draw the concept of the imagination out of this narrow discipline, or rather expose the way in which the imagination naturally extends beyond the confines of the strictly aesthetic. The story of the imagination in this sense leads to epistemology and the philosophy of mind. German Enlightenment thinkers occasionally highlighted what became second nature for pragmatists like Peirce—that the process of the imagination underpins the workings of human cognition. Finally, Peirce sees clearly what Kant begins to reveal. To investigate the imagination is to examine the nature of consciousness and, ultimately, to investigate the character of nature itself. This last move, from epistemology to ontology, is a difficult one to make, but it is facilitated by the later work of both Kant and Peirce.

The transition between the aesthetic, the epistemological, and the ontological will be drawn out once again in the following section on the imagination and the empirical sciences. The section opens with a description of metaphor and image schemas. Perhaps this seems like an odd place to start, but it is in metaphor that we find an instructive case that demonstrates the way that aesthetic imagination structures and gives rise to epistemic claims. Metaphor studies—spearheaded by Johnson and Lakoff—suggest the ubiquitous role that analogy and metaphor play in the creation of human concepts and meaning. As linguists and philosophers have recently proposed, imaginative metaphors serve as the glue of human language and cognition. This work cuts in another important direction to the extent that it suggests that the development of effective metaphor and meaningful analogies relies, almost exclusively, on the physical embodiment of human beings. Humans come by imaginative meaning only in their organic, spatial, and temporal relations with the world. The organism-environment coupling, long studied in the fields of biology, ecology, behavioral psychology, and phenomenology, grants the emergence of stable and meaningful, flexible and innovative concepts. This coupling is, in itself, imaginative in its unique dynamic.

While the following discussion begins by addressing the field of metaphor studies in order to underscore the aesthetic dimension of abstract conceptualization and reasoning, I draw on biology and cognitive neuroscience to investigate the imaginative character of the workings of the human organism acting in its environment. Metaphor studies point to, and continue to support, a theory of embodied cognition. The forthcoming discussion not only provides evidence that imaginative metaphor has its basis in the biological structuring of a given human organism but also suggests that *any* given organism has its vital basis in the force and dynamics of the imagination as hitherto described. This is the move that I have argued Kant makes in the later sections of the third *Critique* and in the *Opus Postumum*. It is a step that Peirce takes in articles such as "Evolutionary Love." It is a claim that I too will make, one that arguably follows from the recent work of the empirical sciences.

Metaphor, Image Schemas, and the Embodiment of the Imagination

In *Metaphors We Live By* (1980), George Lakoff and Mark Johnson first argued for the central role of metaphor in cognition.[2] They suggest that metaphors do not only make our ideas more poetic, beautiful, and interesting but that they provide the very structure of large parts of human conceptualization. Metaphors, often framed as exclusively aesthetic constructs, lie at the foundation of human abstraction and have meaning only to the extent that they appropriate the meanings and relations of bodily experience. In *Philosophy in the Flesh* (1999), they analyze the basic metaphors for such concepts as causation, mind, knowledge, importance, desire, and affection, which are conceptualized in terms of immediate bodily experience. Hence, their study of metaphor naturally led them to the empirical studies of human embodiment emerging from contemporary psychology and cognitive science.[3] Embodied experience— characterized by spatial, temporal, and visceral relations—grounds human abstract conceptualization, which has often been described as highly esoteric and ideal in nature. By suggesting that bodily experience has a foundational role in abstraction, their work begins to question the belief that abstract concepts are produced by allegedly disembodied mind or thought processes. Bodily relations and activities that give meaning to highly abstract concepts are learned over a period of time; they are learned "by rote," literally by the heart, in the flesh and blood of human pursuits.

The emergence of abstract conceptualization depends in large part on the ability to think metaphorically, that is, to project figuratively one set of experiences (in this case, the field of embodied, physical forces and relations) onto a set of experiences of another type. As Johnson observes, "metaphor consists in the projection of structure from one domain onto another domain of a different kind."[4] This projection or mapping occurs between what is termed the "source domain" of immediate experience and the "target domain" of abstract conceptualization. The structured relations and correlations of immediate experience are redeployed in making *sense* of abstract concepts.

This point is made clear by a linguistic example of a complex metaphorical mapping. As many philosophers have discovered, the process of understanding is one of those abstract concepts that is hard to examine. Understanding is difficult to understand. In light of this difficulty, Lakoff and Johnson observed that human beings tend to employ a limited number of metaphors in discussing the conception of mind. One of these is the metaphor of UNDERSTANDING IS SEEING. If I want to convey that I understand a given concept, it is common for me to say, "I *see* what you mean." The experience of seeing objects is mapped onto the way that individuals understand concepts since both vision and intellection depend on the ability to discriminate features and details. Plato, Augustine, and Descartes extend this metaphor in their studies of epistemology—the study of understanding. Their work illustrates the submappings of this basic metaphor. Plato's allegory of the cave is a metaphorical description of the process of coming to understanding. His example demonstrates several submappings of the basic metaphor of UNDERSTANDING IS SEEING. In the cave, individuals can only see shadows of ideas, which Plato describes as things that might be seen more easily under better conditions. Understanding is realized only when individuals are led out of the cave and into the light of day so that people may observe things clearly. In his *Confessions* and *City of God*, Augustine describes the possibility of understanding in terms of the ambient light that God provides. To not understand is to "be in the dark." In *Philosophy in the Flesh*, Lakoff and Johnson observe that, when Descartes appropriated the UNDERSTANDING IS SEEING metaphor, he thereby accepted an ontology of the mental realm that required mental counterparts to visual objects, people who see, natural light sources, and so forth.[5] The ubiquity of such metaphors seems to indicate that even the most esoteric notions can be structured around the relations of immediate bodily experience. Echoing Lakoff and Johnson's observations, Donald Dryden recently wrote, "in domains where there is no discernible pre-conceptual structure to our experience, we spontaneously import structure via metaphorical mappings that ultimately derive their meaningfulness from their ability to match up with pre-conceptual structures."[6]

While the use of body-based metaphors began to point to an embodied theory of cognition that suggests that immediate corporeal experience grounds abstract conceptualization, more work was necessary to expose the way in which this experience is structured and how it might provide a relatively stable framework for conceptualization. Inspired in large part by Kant's work in the first *Critique*, Lakoff and Johnson claim that abstract meaning and highly complex forms of rationality often emerge from, and are limited by, the patterned relations that organize the embodied and emotional lives of human beings. Appropriating and extending Kant's writings on the schematism, Johnson explains that metaphoric projection and its corresponding meaning making depends on what he terms an "image schema." He is careful not to confuse image schemas with "some allegedly pure form-making capacity . . . [or] abstract knowledge structures."[7] Instead, Johnson writes that image schemas should be regarded as "recurring patterns of our sensory motor experience by means of which we can make sense of that experience and reason about it, and that can also be recruited to structure abstract concepts and to carry out inferences about abstract domains of thought."[8]

While discussions of the neurological basis of image schemas often focus on the cross-modality of particular brain areas, it is important to remember that schemas do not arise strictly in the brain. Organisms get on with abstract conceptualization in accord with a particular physiological comportment situated in particular environmental conditions. The patterns of image schemas, Johnson suggests, "emerge as meaningful structures for us chiefly at the level of our bodily movement through space, our manipulation of objects and our perceptual interactions."[9]

This point is made clear in *The Body in the Mind* when Johnson explains the way that the spatial-temporal orientation that constitutes our ordinary experience gives rise to meaningful schematic structures. For example, in the case of the subjective judgment of quantity, humans employ, or "import," the sensory-motor domain of verticality in conceptualizing judgments concerning amount. Let us consider the metaphor MORE IS UP.[10] Here an increase or decrease in quantity and quality is framed by the primary experience of rising and falling respectively. The primary mapping is reflected in various expressions: "Stock prices are *up* and unemployment is *down*." There are also submappings that convey the sense of vertical movement: "Interest rates *soared* and mutual funds *crashed*." But why and how is scale understood in terms of vertical orientation? Johnson suggests the following:

> There are certain basic correlations of structures in our experience that give rise to metaphorical projections of this sort. When we add more of a substance to a container, the level rises. This particular metaphor is not based on similarity, since there are no relevant similarities between MORE and UP. Instead, it is based on a correlation in our experience, of the sort just mentioned.[11]

He suggests that the MORE IS UP metaphor is an instance of what he calls the schema of scale; this schema, like all schemas, is a repeated and engrained pattern of experience. Johnson explains that our conception of scalarity is determined by our immediate experience of quantitative amount and qualitative degree. Like our experience of amount and degree, the scale schema has a fixed directionality, a cumulative and normative character, and can be either open or closed (scale can continue indefinitely or end at a definite point).

It is worthwhile to spend a bit more time examining schematic orientations. All of these orientations are meaningful to the extent that they ground the bodily experience of individuals situated in particular environmental contexts. In each case, these orientations create the bodily foundations of schemas that can be used in various forms of abstraction. Consider the following phrases that employ the orientation of *in-out*, an example that Susan Linder and Mark Johnson develop in some detail in their respective works. Linder conceives of three basic image schemata that employ the *out* orientation. Here I will concentrate on only one of the three. Consider the following phrases:

Tracy went *out* of the building.

Cut *out* that section of tape.

Pick *out* your favorite piece of music.

Trace *out* the silhouette of the face.

In each case, the expression *out* is being employed and conveys the same basic orientation that makes the respective phrases meaningful. "Out" is understood relative to a "CONTAINER schema" in which there is a definite boundary, an interior and an exterior. Such orientations are encountered by humans every day in their immediate experience. Elaborating on this CONTAINER schema, Linder explains that the meaning of "out" depends on the basic relation of a "landmark" that remains stationary relative to a "trajector" that moves.[12] For instance, in the above example, Tracy would be regarded as the trajector while the building serves as the landmark. "The relevant schema," Johnson writes, "is the repeatable pattern of out movement in each of the specific actions. Notice that in each different case the schema is realized in a different way, though it retains a recognizable form."[13] This point underscores the flexible stability of image-schematic thinking, a stable flexibility that has been drawn out in our discussion of the imagination. Image schemas are stable insofar as they depend on the particular comportment of an organism and rely on the patterned ways in which an organism experiences itself in the world. If image schemas change at all, they change in accord with evolutionary development. That is to say, they change *very* slowly. That being said, image-schematic thinking demonstrates a creative and flexible process by which basic experiential relations can be redeployed and reapplied in interpreting new conceptual data and making new expressions. As Dewey pointed out, when the old is made anew in experience, there is imagination. I will revisit this point shortly in the upcoming discussion of prototypes and radial categories.

The process of metaphoric mapping that Johnson describes is reminiscent of Kant's comments in the "Dialectic of Aesthetic Judgment" in the third *Critique* and Peirce's comments on the character of mediation and abduction. Kant notes that the bodily experience associated with words such as "ground" (basis), "depend" (to hold up from above), and "flow" give meaning to a variety of abstract usages, in a process that he refers to as "symbolic hypotyposis."[14] Kant describes this process in section 59, entitled "Of Beauty as the Symbol of Morality." In his rendering of symbolic hypotyposis, Kant writes, "to a concept only thinkable to reason, to which no sensible intuition can be adequate, an intuition is supplied with which accords a procedure of the judgment analogous to what it observes in the schematism, i.e. merely analogous to the rule of this procedure."[15] Kant does not go into much detail to explain the procedure that operates by way of analogy, but he does seem to intimate that this is a creative process that does not depend on forms of determinate judgment. It is worth emphasizing that a metaphoric projection is a reflective process of mapping disparate fields of human experience that harmonize—and harmonize even in the absence of preestablished rules. Another way of saying this is that the premises and domains of imaginative consciousness cannot be pregiven or exhaustively described prior to employment of the mapping.

The mediating process that acts to draw seemingly separate maps into harmony has been described by Lakoff and Johnson as being "imaginative" insofar as it is defined by a particular mediation, plasticity, and variability. While reflecting these three imaginative characteristics, metaphoric mapping provides a relatively stable framework for conventional conceptual metaphors. In his work on Peirce and metaphor, Carl Hausman emphasizes the creative rather than the conventional aspect of metaphor, commenting that metaphor, like abduction, ought to be associated with novel forms of expression and thought. He notes that "one of the most puzzling things that metaphors seem to do is articulate new insights; they create what they discover."[16] Hausman's work, in conjunction with that of Lakoff and Johnson, indicates that our interest in imaginative metaphor should not be confined to a particular study of aesthetics. Instead, these authors intimate that aesthetics as a whole—and the flexibility and creativity that have been associated with it—may serve a central function in the development of human cognition.

Flexible Stability: Conceptualization, Categorization, and Prototypes

Their investigation of metaphor led Lakoff and Johnson to identify a type of flexible stability as an important aspect of human understanding. It also led them to suspect that this imaginative characteristic might be reflected in more ordinary forms of conceptualization and categorization. They thus turned their attention to the formation of conceptual categorization by way of prototype construction. They observe that many forms of abstract categorization depend on the existence of prototypes, that is, neural structures and neural dynamics that allow human beings to make an inference or do an imaginative task relative to a category.[17]

Before arriving at the neural basis of the prototype, they explored a number of different ways of defining the term more generally. At first, they suggested that a prototype is a member of category that serves as the typical instance of the category, a central benchmark against which other noncentral members are judged. A prototype is not given a priori but is entirely dependent on a given interaction between an agent of inquiry and its environment. As Eleanor Rosch noted in the 1970s, the prototype theory of conceptualization stands as a departure from the Aristotelian conception of the categories as being established by necessary and sufficient conditions.[18] Instead of a definitional theory of categories that operates by way of such conditions, Rosch suggests that individuals come to establish and accept the central members of a category over time and in particular experiential circumstances. A prototype is established and modified by the exposure to particular experiential cases but also sets the stage for the interpretation of future experiential instances. The "distance" from a prototype to a particular member case would be a function of its "closeness" to the central prototype—a function of shared qualities, correlations, or, as Lakoff underscores, metaphoric-metonymic extensions.

Although Rosch identifies the existence of conceptual prototypes, she does not address the way that prototypes might be used to provide category structure. Lakoff seems to recognize this point and elaborates on the basic understanding of prototype effects in order to develop radial category structures. A radial category is one in which members of the category are extended from a central case by means of particular principles. In 1987, Lakoff identifies the Japanese word *hon* as defining a type of complex radial category. Central cases of *hon* convey the meaning of "a thin long object." The noncentral members of this category include those activities done with *long thin* objects (sword contests), communications (because of the *long thin* wires used), trajectories (*long thin* lines established by the movement of an object), and activities that are *like* activities done with long thin objects (judo and koan competitions).[19] Notice that these cases extend from the central example that unifies the various meanings and that these meanings can radiate from the central member by way of metonymic and metaphoric constructions. The principles of extension employ a function, definition, or analogy of the central conceptual reference point.

When Claudia Brugman examined the prepositional semantics of "over," for example, she employed radial category structure and underlined its metaphorical extensions. In the central case, "over" connotes the meaning of "above and across." While certain extensions are concrete and can be traced to the function or definition of the central case, such as the extension of "walking over the mountain," others are metaphorical and convey the meaning of repetition (to do something over) or excess (to overwork or overdo). In these cases, the central case serves as a type of source for the meaning of the metaphoric extension.[20] Brugman's work shows how the source and target domains of metaphor studies might be employed to explain particular radial category structures. More generally, Brugman and Lakoff demonstrate that the same linguistic proposition can have the ability to convey multiple meanings that are related by means of a common reference point. In the case of "over," the term can convey the meaning of excess, repetition, and literal spatial orientation depending on the linguistic and situational context in which it is expressed. This also means that the same proposition can be applied to different contexts while maintaining its basic schematic or prototypic relation. This type of economy and flexibility has been discussed earlier in reference to the standards of abductive insight and the nature of the imagination.

Keeping these comments about flexible conceptualization in mind, it is necessary to underline that prototypes provide a *relatively* stable conceptual framework to make inferences. This is not to say, however, that conceptual categories cannot change. They surely can. However, conceptual reference points usually change rather slowly in light of new environmental situations or experiential inputs. Occasionally, they shift to accommodate the novel circumstances that show themselves over the course of inquiry. Only very rarely do they change dramatically in order to adapt to the radical alterations that occur in their environments. This being said, in the absence of any "special contextual information," prototypes provide us with one of the important heuristic devices that we employ to interpret and filter what William James called the "buzzing blooming confusion of experience." To say that many conceptual categories depend on prototypes is also to say that, in many cases, concept formation is a matter of degree and correlation, operating by way of analogy-metaphor or relative similarity and difference. These points resonate with Hebb's rule of neural plasticity, which will be examined in coming sections. The rule will be addressed again at the end of the following chapter, when I introduce the work of Peter Gardenfors, work that points to the flexible and experiential character of category construction. We will see in the coming sections how neural architecture may grant the possibility of metaphoric and prototypic flexibility.

Stuart Kauffman's analysis of organic complexity will be employed in later sections to elaborate on this point and to suggest that heuristic prototypes and human metaphor are rooted in primitive, organic forms of mediation, plasticity, and spontaneity. Speaking somewhat loosely, Kauffman notes that in the face of a biosphere of nearly infinite variation and complexity, embodied metaphor and prototype heuristics might be the only way for an organism to "get on with his/her business." To put this point somewhat differently, if the propositional content of our language and conceptualization do not have a one-to-one relation with the particularity of experiential situations—situations that are continually transforming—there must be another way of generalizing and making sense of this experience. Image schemas and prototypes provide human beings a type of heuristic shorthand that allows them to negotiate complexity by way of relatively simple and stable constructs. Kauffman writes: "And metaphor? If we cannot *deduce* it all, if the biosphere's ramblings are richer than the algorithmic, then metaphor must be part of our cognitive capacity to guide action in the absence of deduction."[21]

An important point needs to be made before turning to the bodily basis of metaphoric thinking. Most of Lakoff and Johnson's early work aimed to demonstrate the ubiquity and "fact" of metaphor. In essence, they showed *that* metaphoric mapping is at the core of abstract judgment. More recently, they have become interested in the more ambitious project of exposing *how* metaphoric mapping and experiential-abstract conflation function in human judgment. The coming sections concentrate on this latter project. I am interested in the way that this mapping occurs and is rooted in physical processes that reflect the principles of spontaneity, growth, and mediation typically associated with the aesthetic imagination.

In the midst of writing a second draft of "Thinking and Cerebration" in 1880, Peirce scratched out a large swath of writing. Thankfully, the words can still be seen: "The connection of the mind with the nervous system is so intimate that the essential laws of the former must necessary [*sic*] correspond to those of the latter."[22] In situating embodiment studies at the heart of epistemology, Peirce expresses a sentiment that has for too long been scratched out and omitted in the history of Western thought. Indeed, embodiment studies only came into their own with the flourishing of phenomenology in Europe in the early twentieth century and only began to hit their stride in the late 1970s, with the debates surrounding the field of cognitive science. Even then, embodied and imaginative theories of language and cognition were still considered to be the weak cousins of more analytically rigorous propositional accounts.[23] Only very recently have these theoretical underdogs been vindicated and recognized for their foresight in situating the embodied and triadic (metaphoric) imagination at the core of human cognition.

Coming to Our Senses: Affect and the Body of Thought

In recent years, convergent evidence from the cognitive neurosciences has pointed to the neural and bodily basis of metaphor and suggested that image schemata ought to be considered "dynamic activation patterns that are shared across the neural maps of the sensory motor cortex."[24] More plainly, the evidence has begun to show that the brain is fundamentally multimodal and cross-modal. Johnson's more recent work in the *Meaning of the Body* (2008) focuses on this multimodal processing and neural differentiation that grants the possibility of metaphorical thinking.[25] I will consider this evidence first by way of broad strokes and then with more detailed accounts of neural development, architecture, and function that begin to describe the embodied basis of imaginative thought.

Lakoff repeatedly underscores the multimodality of actions, highlighting the way in which motor, perceptual, and somatosensory components are coordinated. For example, these components allow an individual respectively to *do* an action, to *perceive* the action being done, and to "get the sense" of doing the action. This coordination is reflected in neural activation patterns, the study of which gave rise to the hypothesis that multimodal coordination might ground abstract thought. After exploring this hypothesis, researchers found that there is a simultaneous coordination of different neural domains that underpin the mapping between the metaphoric domains that Lakoff and Johnson began to identify in the 1980s. Specifically, recent work has indicated that there is a continual coordination between the sensory-motor domains and the neural domains that have long been regarded as the seat of abstract conceptualization. This neural multimodality has come to the fore in the study of cognitive linguistics.

Debunking the long-held position that Broca's and Wernicke's areas were the exclusive neural loci of semantic understanding and language production, studies have shown that the sensorimotor cortices are crucial to semantic understanding of bodily action terms and sentences.[26] In his recent meta-analysis of metaphoric cognition, Timothy Rohrer employs contemporary fMRI and ERP experiments to highlight the way literal and metaphoric stimuli activate areas in the sensorimotor cortex that are consistent with the image schemata hypothesis. In these studies Rohrer first attempts to show that brain areas researchers once assumed were only activated by spatial and bodily orientations are also activated by linguistic cues that describe these particular orientations. The neural activation detected when one picks up a box is largely isomorphic with the activation stimulated by the command to "pick up that box." Second, and more significantly for our study of metaphor and image schema, Rohrer discovers that describing particular physical orientations serves as a literal cue that generates activation patterns largely isomorphic with the patterns detected when a subject is exposed to polysemous cues, that is, cues or signs with multiple meanings. For example, the literal expression "pick that box up" activates the same neural pattern as the metaphoric or polysemic expressions such as "turn up the volume" or "he turned it up in that basketball game." This result obtained when other schemas, such as the "out" schema discussed earlier, were tested in a similar way. In short, polysemous cues trigger spatial and corporeal relations, indicating that the "mind" that handles abstract conceptualization can, in an important sense, be found in the bodily relations of a human organism. This is not merely to make the claim that one needs a body to think but rather the stronger claim that our bodies, and their relationship with their environmental situations, continually structure human thinking.

Rohrer notes that image schemas and corresponding neural maps develop over time, evolving to accommodate novel situations, and, in this development, actually exhibit a type of emergent creativity of their own. The work of Rohrer indicates that abstract conceptualization appropriated or co-opted the structured neural relations of the sensorimotor cortex in the development of the brain architecture that could support abstract cognition.

Gerald Edelman's concept of secondary neural repertoires echoes Rohrer's account, and it is a neural process that likely explains how integrative areas of the sensorimotor cortex acquire both sensorimotor and image-schematic functions.[27] Edelman's work will be instructive in later sections of this book, in which the investigation of the imagination will lead to an inquiry of biological organization and growth. For now, a brief description of Edelman's account will outline the degree to which this imaginative structuring—in its emergent character—obtains in the physical activities of a living organism. More simply, the architecture and dynamics of the human nervous system is continuous with, and continually structures, the life of the mind. In his description of secondary neural repertoires, Edelman helps us explain the possibility of the metaphoric process of the imagination, the mapping of particular image schemas, in highly specific physical processes that demonstrate the similar modes of organization and emergence.

Edelman argues that neuroembryonic development (the development of neural maps) can best be understood as a process he calls neural Darwinism. First, through the processes of cell division, growth, and selection, neural sites are established. These sites should be regarded as local neural regions that possess particular patterns of dendritic and axonal arborization determined by morphoregulatory molecules that affect the neural architecture of these particular regions. This developmental selection produces what Edelman calls "primary repertoires" consisting of large numbers of variant neural circuits within particular anatomical regions.

As organic agents, neurons in the embryo seem to flourish and find nourishment and particular forms of reinforcement—first in developmental selection, then in experiential selection. Neuronal synapses are the units of selection in this model. They undergo and participate in a process of neural amplification. In so doing, neural activation patterns make the rather curious journey from chaos to order. As Edelman highlights, this is a form of experiential selection that depends on environmental conditions and the behavior of the organism. Under certain conditions and in light of certain behaviors, some neural networks will be activated more than others. The differential activation of neuronal selection serves the same function that differential reproduction serves in natural selection, creating the conditions by which certain networks actively thrive even as others become "extinct" or are "pruned." Over time, the neurons that activate in tandem become physically correlated, developing Hebbian associations that engrain and reinforce particular patterns of neural activity. In an overused expression, "neurons that fire together wire together," creating functional clusters that mediate and coordinate the activity of an array of neurons. Through this process, genuine Peircean "thirds" emerge, that is, functional entities irreducible to the two parts between which they meditate. Hebbian association can be described by the function given below, a function Peirce roughly approximates in his discussions of human physiology.

Seventy years later, in 1949, Hebb wrote *The Organization of Behavior*, in which he postulated: "When an axon of cell A is near enough to excite a cell B and repeatedly or persistently takes part in firing it, some growth process or metabolic change takes place in one or both cells such that A's efficiency, as one of the cells firing B, is increased."[28] According to Hebb, this correlation can be mathematically described in the following manner:

$$W_{ij} = X_i X_j$$

W_{ij} is the weight of the connection from neuron j to i, and X_i is the input for neuron i. This equation can be rewritten in the following way, which will allow us to analyze its strengths and shortcomings:

$$\tau_w \frac{d\mathbf{w}}{dt} = v\mathbf{u}, \quad \mathbf{w}_{t+1} = \mathbf{w}_t + \frac{\delta t}{\tau_w} v\mathbf{u}$$

This hypothesis, borne out in modern cognitive science, is consistent with Peirce's understanding of habit formation but also points to the way in which novel input stimuli can revise these habits of association. This formula, however, is misleading in its simplicity. In the years after Hebb's discovery, scholars identified two problems with Hebb's theory. First, Hebb describes no process by which connections can get *weaker* over time. Second, Hebb's rule provides no upper bound or limit for how strong connections can get. This is to say that Hebb's rule was unstable. And this is where what appears easy at first gets very, very hard. Bienenstock, Cooper, and Munro proposed a model (BCM) of synaptic modification that holds that neurons possess a synaptic modification threshold that is not fixed but instead varies according to a nonlinear function with the average output of the cell:[29]

$$\tau_w \frac{d\mathbf{w}}{dt} = v(v - \theta_v)\mathbf{u}$$

$$\tau_{\theta_v} \frac{d\theta_v}{dt} = v^2 - \theta_v$$

$$\tau_w < \tau_{\theta_v}$$

More recently, Byrne translated this postulate into a variety of quantitative expressions in which a neuron A with average firing weight V_A projects to neuron B with average firing weight V_B. The synaptic connection from A to B has strength T_{AB}, which determines the degree of activity in A is capable of exciting in B; the strength of T_{AB} should be modified in some way that is dependent on *both* the activity of A and B. The general expression of this plasticity rule is formulated by the function:[30]

$$\Delta T_{AB} = F(V_A, V_B)$$

With the development of this function, Hebb lays the groundwork for the connectionist theorists of the 1960s, who began to create models that envision brain functions as emergent with global properties that result from the interaction of connected neural networks. Hebb's rule is also significant in the sense that it identifies growth and metabolic change as the defining features of our nervous system, features that coincide with the spontaneity and creativity that have been associated with the imagination. Several points, however, need to be highlighted to recognize and qualify the import of Hebb's insight.

First, the metabolic growth (ΔT_{AB}) is calculated as a function of the instantaneous interaction of two individual cells. It is the case that growth depends on the actions and reactions of *both* cells and emerges as a process that is irreducible to the excitation of either A or B in isolation. What some interpretations of Hebb's rule neglect, however, in their attempts to remain parsimonious, is that these two cells are *already* interacting in wider neuronal complexes that set the limits and provide the enabling conditions of the cells' excitation. This limit and enablement arises in the ongoing and nonrepeatable transformation of brain states and metabolic processes. Edelman points to these processes in his description of the creation of secondary repertoires. In emphasizing this point, we can see that Hebb's rule cannot be easily applied to the activation dynamics of neural populations, since these complex dynamics cannot be modeled by a mechanical rule or linear function. I will return to this point in my discussion of neural reentry.

A second point needs to be emphasized in regard to Hebb's rule: the metabolic growth rate is calculated as change at a particular instant in time. This calculation provides only a snapshot or glimpse of neural dynamics, without demonstrating the course of events and relations that might have led to this rate of change. That is also to say that Hebb's rule provides an idealized model of an extremely small time interval of continuous and diverse evolutionary, developmental, and experiential processes. This is not to dismiss Hebb's work but to encourage us to extend and modify the rule better to demonstrate the continuity of neural activation and development.

This qualification of Hebb's rule suggests that neural dynamics should not be regarded as mechanical linear functions. This said, the rule also does not support the idea that these dynamics are the products of random chance or pure contingency. To return to a point made earlier, *this form of creative irregularity is irreversible and directed.* As the psychiatrist Jeffrey Schwartz notes in his experiments on language acquisition, "Once the Hebbian process has claimed circuits, they are hard-wired for that sound; so far, neuroscientists have not observed any situations in which the Hebbian process is reversed so that someone born into a sea of, say, Finnish loses the ability to hear Finnish unique sounds."[31] The metabolic change that occurs is an irrevocable fact that continues to affect—albeit in a dissipating or decaying degree—the future transformations in the neural activation of the system. It is in this sense that neural dynamics are motivated by and arise in the creative spontaneity that has been outlined in our discussion of the imagination and in Peirce's description of abduction and tychism.

It is worth noting that Peirce makes an identical point in "Thinking and Cerebration" in 1879: "Neural activation causes fatigue, but long continued stimulation causes another phenomenon, namely, the spread from cell to cell of the nervous activity . . . along whatever path a nervous discharge takes place, along that path a new discharge is likely to take place."[32] This spread of activation causes physical and dynamic adjustments in our neuronal complexes over time that are genuinely novel and irreversible. William James, following in Peirce's footsteps, anticipates

Hebb's rule when he writes on the process of mental association in his *Psychology* in 1890: "The amount of activity at any one point in the brain cortex is the sum of the tendencies of all other points to discharge into it. . . . When two brain processes have been active together in immediate succession, one of them on recurring, tends to propagate the other."[33] In his recent studies, Rohrer elaborates on this point, explaining that neurons aggregate over time in neuronal groups in a process that can be described by rules of plasticity but, of course, cannot be predicted or anticipated by these rules. These aggregates in turn "act like organisms that seek out stimulation as nourishment, and the neuronal groups compete with each other as they migrate along the neural tube toward the emerging sense organs."[34] Hebb's rule describes this aggregation and lends credence to the insights of Peirce and James.

The migration and population-growth dynamics of the neuronal groups create yet another emergent property: neurons array themselves into physical patterns that "map" the various sensory modalities. This mapping refers to the fact, culled from an array of experiments since the 1960s, that particular posterior regions of the brain space are associated with particular sensory functions. More carefully stated, Edelman explains that the use of physical space in the posterior regions of the brain to represent environmental stimuli provides the incipient topographical neural maps of the sensory modalities. To reference topology is appropriate in this case since Gallistel and others have observed that the vector spaces of topology have a literal interpretation in the nervous system.[35]

In phase 1 of neuronal group selection, developmental selection occurs as a result of growth-factor signaling and selective pressures that "yield anatomical networks in each individual."[36] This involves the development of neural sites described earlier and provides a primary neural repertoire. In phase 2, "selective strengthening or weakening of particular populations of synapses occurs as a result of behavior."[37] The weighting of synaptic activation creates a relatively stable organic framework that underpins our relatively stable mental lives. It is worth noting that the Darwinian concepts of propagation, selection, and adaptation—concepts that underpin evolutionary theory—have been imported into Edelman's account of physiological development. It is for this reason that

Edelman's theory is often referred to as somatic evolution. Finally, in phase 3, reentry, a process that we will examine in some detail in the following chapter, coordinates neural maps through "the parallel selection and correlation of neuronal groups in different areas" of the brain, making possible the emergence of complex sensory and conceptual meanings.[38] After multiple exposures to a stimulus, activation patterns are established in a variety of mapped areas. "Operations in these different maps that are related to the same perceptual stimulus are linked together by reentry."[39] An example is helpful in elucidating this last stage of neuronal organization. As Donald Tucker notes, "The primary auditory cortex is mapped by frequency (pitch of the sounds, high or low), not by higher perceptual objects such as words. Therefore, processing at deeper or 'higher' levels is required to form the perception of a word that gets meaning from the sounds." The comprehension of a word depends upon heteromodal coordination involving numerous functional maps. This heteromodal coordination and synchronization is what Edelman refers to as reentry.

Cross-modal coordination grants the possibility of the categorization and abstract conceptualization that Rohrer identified in his fMRI and ERP experiments described earlier. In terms of the investigation of image-schematic and metaphoric forms of cognition, it is worth mentioning that auditory areas develop maps indicating increasing pitch and volume in this way. Later, as Rohrer notes, "tactile areas develop somatic maps for pain and touch along limbs; and still later, somatomotor maps develop for muscles distributed across the limbs."[40] The schematic of neural adaptation and selection is deceptive in the sense that the level of analysis shifts between the interneural dynamics of phase 1 and phase 2 and the intermap dynamics of reentry depicted in phase 3 of the schematic. This shift is confusing but necessary in order to present the emergent phenomenon of reentry between developing neural regions.

Rohrer hypothesizes that the synchronization of neural maps is necessary for the establishment of cross-modal image-schematic patterns. This is most notable between auditory and tactile neural nodes. It is worth pointing out that the development and coordination of neural maps does not stop in any determinate manner. Indeed, this neural development, with its simultaneous propagation of connectivity and diversity, proceeds in a way that cannot be predicted nor exhaustively described, exhibiting a type of plasticity that has been described as uniquely imaginative.[41]

This fact is borne out in several studies. For example, in their experiments with primates, Allard and colleagues have demonstrated that organisms are free to reorganize dynamically the somatosensory cortical maps within certain constraints.[42] Areas lacking their previous sensory connections were "colonized" in a couple of weeks by adjacent neural maps with active connections. Experiments conducted in the 1980s found that activation patterns in the cortex that mapped sensory input from fingers underwent a distinct and orderly shift when a finger was removed. When the middle finger was removed, the neural spaces designated for the ring and index finger would enlarge, taking over the dysfunctional middle-finger map, in effect compensating for the loss of the finger.[43] What is interesting about these studies is that this physiological "colonization," the process by which latent neural structures are appropriated and new aspects of brain architecture are utilized, coincides with a type of behavioral novelty, the ability to conduct new activities in light of unprecedented environmental circumstances. Freedom-within-constraint, that odd disposition of Kantian imagination and Peircean inquiry, shows itself in dynamic neural development. This fact may indicate that the novelty of experience and inquiry are grounded in the architecture and function of the neural circuitry of our nervous systems.

In the growth and development of multimodal processes, the neural maps continue to adapt and "learn," taking advantage of the latent organizational possibilities of their constitutive systems. This process of "learning" is unique to the particular physical structures of individual brains. Imaginative originality shows itself at multiple levels of analysis: organisms perform new and imaginative functions by virtue of, and in tandem with, the novel neuronal organization that obtains in their physical embodiment. As Jerome Feldman recently wrote, "mental connections *are* active neural connections."[44] At first glance, this point may seem to be overstated in at least one respect: mental connections are accompanied by a particular quality of feeling that cannot be reduced to the quantitative study of neural activation. Feldman admits that "the pleasure of beauty, the pain of disappointment, and even the feeling of being alive do not seem to us like they are reducible to neural firings and chemical reactions."[45]

I take Feldman's main point, however, to be that mental connections depend upon and emerge from neural activation. Mind is an aspect of biological and neurophysiological rhythm. A recognition of this dependence will force us to revise our epistemic and metaphysical assumptions. Feldman's comment encourages us to amend at least two longstanding epistemological commitments—our commitment to Cartesian mentalism and our common understanding of materialism. First, in terms of revising the effects of our Cartesian legacy, the discoveries associated with neuronal organization suggest that our imaginative lives are inextricably connected to our embodied lives, to the patterned relations that obtain in our physical makeup. Second, what we discover upon a close investigation of this makeup is that the spontaneity, mediation, and plasticity that we have historically associated with the creation of fine art and refined thought are demonstrated at the level of material organization and processes. It is in this sense that the current study of cognitive neuroscience seeks to amend the doctrine of materialism.

Imaginative Adaptation and Mirror Neurons

The architecture of the brain, while providing an enormous degree of variability and possibility, sets constraints on the development of certain activation patterns. The structure of the brain, in effect, sets the stage for future activation that is both free and constrained. Just as pragmatic inquiry adjusts its scope and direction over time within determinate constraints, the size and boundaries of neural maps can be changed in light of novel environmental conditions, experience-dependent learning, and social interactions. This similarity between pragmatic inquiry and neuronal adaptation is not incidental but rather makes sense of Peirce's claim that the development and growth of thinking must arise from the development of cerebration. To make the same point in a different key, researchers are beginning to identify the physiological basis of the adaptive thinking and imaginative coordination that has so often been described phenomenologically in the accounts of the classical American pragmatists.

In the late 1990s, Vittorio Gallese, Giacomo Rizzolatti, and others began to investigate what would later be called the mirror neuron system. Their main goal was to expose the neural mechanisms by which primates and human beings might understand and imitate particular actions. As Umilta and colleagues summarize, the mirror neuron hypothesis asserts that there must be neural systems that recognize the actions of others. This recognition is achieved by matching the observed action on neurons that motorically coded the same action. By means of such a neural matching system, the observer during action observation is placed in the same "internal" situation as when actively activating the same action.[46]

Exploring this hypothesis, researchers identified a set of neurons in the premotor cortex of humans and some primates that provide the capacity for a nearly instantaneous response on an unconscious level *both* to external and internal cues. Through a mapping of particular brain areas, researchers discovered nerve nets that were activated *both* by the subject's observation of a meaningful action *and* by the actual performance of the action.[47] For example, the same neuronal activation occurs when one grips the handle of a hammer and when one sees another person gripping a hammer in a similar way. It could be said that the social realm in which emotions are embodied and action takes place very literally gets under one's skin. In a colloquial sense, neuronal activation does not make a distinction between the actions and intentions of another and the actions and intentions of oneself.

The crucial point here is that these neural nets are unique in their ability to respond to, be activated by, *and anticipate* what comes next through the subject's observations of complicated procedures. Again, to speak loosely, the neurons anticipate and react to the agency of others just as they would anticipate the agency of oneself. These findings are important for many reasons, but perhaps most notably in the way that they revise the standard understanding of neurophysiology and neural activity. Neural activity should not be regarded as a mechanical process delimited by a particular skull, enclosed in a particular black box. Such activity is always already "out there"—that is to say, always already in the world—responding, coordinating, *mirroring* the dynamics of a natural environment. At first glance, it might seem that the evidence indicates that this mirroring occurs only between animate agents or social individuals. This is only partially true. It should be remembered that these agents are a part of nature and, indeed, are constituted by its dynamic processes. It is in this sense that the mirror neuron system allows us to mirror our natural environment.

Three additional points need to be made regarding the mirror neuron system. These points are, at the very least, suggestive to our discussion of the imagination and human cognition.

1. Gallese's work on the mirror neuron system indicates that similar neural activity is found in human beings when they *perform* an action as well as when they *imagine* or *think* about doing that selfsame action. This fact seems to point to the neurophysiological basis of learning by way of experiential priming. Interestingly, the visual stimuli most effective in triggering these mirror neurons were the subject's observations of actions "in which the experimenter's hand or mouth interacted with objects."[48] From an evolutionary perspective, this should come as no real surprise; the mouth and hands are obviously crucial in the acquisition of food and integral to the sociality of most mammals.

2. Recent studies conducted by Kohler and colleagues demonstrate that neurons in the ventral premotor cortex (F5) of macaque monkeys fire and are suppressed both when the animal performs a specific action and when it hears a related sound. Most of the neural nets also discharge when one observes or hears the actions of another organism performing this activity.[49] This work indicates that partial stimuli have the ability to cause more general neural outputs. The *sound* of a peanut cracking generates the neural outputs that occur when an animal performs the action of cracking a peanut and when the macaque observes another animal cracking the nut. These studies highlight the way in which the understanding of the actions, traditionally framed as intrapersonal and insular, might arise in and through that body's creative interaction with the social-environmental sphere.

3. By extending these studies, Umilta and colleagues hypothesized that neural output associated with the performance of certain actions could be produced in animals by allowing them to observe only a small snippet of that action being performed. Their tests, which employed two basic experimental conditions, supported this hypothesis. In the "full

vision" condition, a macaque was allowed to observe an action directed toward an object. In the "hidden condition," the macaque was allowed to observe the same action, except that in this case the final critical stage of the action (hand-object interaction) was shielded from the subject's view. In both conditions, the output in the mirror neuron system was largely the same, indicating that such neural responses might allow me to, in Umilta's words, "know what you are doing" even under conditions defined by partial information.[50] This point sheds light on the way that bodily comportment and neuronal architecture might grant the possibility of hypothesis formation and resonates closely with Peirce's understanding of theorematic reasoning, which proceeds to a conclusion that is not prefigured in particular premises.

The research on the mirror neuron system is significant in our investigation of the imagination in the sense that it begins to point to a physiological process that allows organisms to be in touch with their local situations, make generalizations from partial observations, and adapt to their particular circumstances in the continuous flow of memory, inquiry, and learning. Our discussion of the imagination has elucidated the way in which imagination allows us to "grasp" and "handle" the novel circumstances that our surroundings afford. Furthermore, we have seen that the imagination as described in the sections on abduction plays a central role in our ability to make new generalizations from partial cases. Consideration of the studies on the mirror neuron systems allows us to explore the physiological basis of these imaginative abilities.

While a description of cross-modal coordination begins to point to the neurology of the imagination, a body of literature has recently extended these descriptions by examining the neural dynamics that seem to exhibit the novelty and adaptability that we have come to expect of the imagination. More specifically, Donald Tucker's work on the "core-shell" model of the brain and Gerald Edelman's investigations of reentrant and degenerate neural dynamics make real headway in surveying the physical ground in which the imagination is rooted. The next chapter concentrates on their respective works.

EIGHT

EMERGENCE, COMPLEXITY, AND CREATIVITY

The Unknown Root of Reason: The Core-Shell Hypothesis

To say that the brain is functionally multimodal begins to point to the organic basis of metaphoric thought.[1] More generally, it points to the way in which affect and bodily sense provide the ground for complex human understanding. As Kant and Peirce both recognize, the mediation between sense and understanding is the domain of the imagination. In this respect, Donald Tucker's *Mind from Body: The Neural Structures of Experience* could be regarded as a treatise on the imaginative basis of human cognition. Tucker's work, unlike the empirical studies mentioned previously, provides a cohesive and detailed account of the way in which neural structures coordinate affect with higher, more differentiated modes of cognition. Tim Rohrer's work demonstrates that specific abstract concepts are grounded in specific sensory-motor experiences, successfully exposing the neural and bodily basis of specific metaphors. Tucker goes a step further. His ambitious project argues that the meaning and fullness of human cognition arises in a continual mediation between visceral and emotive experience supported by the neural connectivity of the limbic system, on the one hand, and the more differentiated sensory and motor centers supported by the neocortical areas in each of the brain's two hemispheres. This is also to say that the complexity and depth of abstract cognition cannot be traced to a particular neural region but develops in the interplay of numerous areas, an interplay that coordinates both visceral sensation and more differentiated modes of activation.

Tucker argues that "the mind comes from the body" in two significant respects. First, he shows that particular patterns of conceptualization arise from the neural activation patterns of various brain areas. Second, he shows that visceral-emotional experiences—often marginalized in philosophical accounts of cognition—ground the processes of abstract conceptualization that help human beings negotiate the world. This two-pronged argument relates directly to our study of the imagination. Tucker regards abstract cognition as growing out of bodily sense and, in Kant's words, as a type of "natural gift," the offshoot of particular organic processes.

So what do these processes look like when examined from an empirical perspective? Such a question has to be answered if, as Tucker suggests, the human mind is "instantiated in a highly patterned differentiated architecture."[2] First, Tucker suggests that the brain exhibits a vertically integrated structure that is common in mammalian evolution. Very simply, the verticality exhibited in animal encephalization suggests that more specialized structures arise from, and co-opt, less differentiated structures. What is distinctive about the human brain, however, is both the degree of complexity and the degree of differentiation that it exhibits. Peirce writes in 1880 that "complexity is almost an essential feature of a nervous organism."[3] Tucker refines and extends this point in his examination of the relation between the human mind and its neural processes and structures.

The brain can be regarded as possessing distinct yet highly interconnected neural levels that vary in terms of the degree of neural differentiation. The brain's dense core, the limbic system, contributes to the determination of visceral control.[4] This core is a *massively* interconnected structure that is found in all vertebrates. Rather surprisingly, Tucker finds that this visceral-emotional control center serves as a type of crossing guard and director of the "neural traffic" of the human brain. This came as a shock to many researchers who believed that the limbic system stood as the remnants of "ancient" neurology. This finding resonates with Peirce's claim that any thought, and more particularly any hypothesis, is shot through by feeling.[5] Tucker reminds us that, in Peirce's words, "there is a peculiar sensation belonging to the act of thinking that each of these predicates inheres in the subject." By the same token, Kant suggests that there might be a "common sense" that grounds aesthetic judgment and that such judgments coincide with a communicable feeling. With regard to the discussion of the relation between sensibility and aesthetic judgment, this is the quality of feeling upon which aesthetic ideas are "hooked." Tucker explores the physiological basis of these speculations in his investigation of the thalamus and limbic system.

Despite the presence of the limbic system in the different stages of vertebrate evolution, Tucker emphasizes that the system is not merely vestigial but rather fully functional in human conceptualization. Conceptualization, memory, and the creation of meaning depend on the interaction and bidirectional mediation between this core and the external shell, a shell which is composed of neocortical somatic control regions and sensory regions (the frontal lobe, motor strip, sensory strip, parietal lobes, and occipital lobes). The corticolimbic pathways link this neocortical shell to the limbic core and exhibit a density of neural interconnections. These two-way neural highways, according to Tucker, provide a way of mediating between sensations culled from an organism's environment (mirrored in the external shell) and the visceral-emotive meaning relayed from the limbic root. Tucker explains that

> at the (limbic) core must be the most integrated concepts, formed through the fusion of many elements through the dense web of interconnection. This fusion of highly processed sensory and motor information (abstracted through three previous levels of network processing), together with direct motivational influences of the hypothalamus, would create a syncretic form of experience. Meaning is rich, deep, with elements fused in a holistic matrix of information, a matrix charged with visceral significance.[6]

The neocortical areas that radiate from the visceral core reflect progressively higher degrees of differentiation, the product of more recent evolutionary adaptation. Relative to the "deeper" processes of the visceral core, these areas are functionally and structurally separate from the general workings of hemisphere operations. Tucker notes that this sort of differentiation is exhibited most pointedly in the cortices' outermost shell, which, rather than being constrained and controlled *primarily* by visceral output, is delimited and determined by sensory input. These outer areas create a topological map of "environmental stimuli that is in turn mirrored in the sensory receptors."[7]

The bidirectionality of neural traffic between the visceral core and the neocortical shell illuminates our discussion of the imagination. In the *Critique of Pure Reason*, Kant suggests that human cognition is achieved only when its emotive-sensuous component is brought under the legislation of the formal understanding. In determinate judgments, emotion is subjected and subordinated to the power of understanding, and the imagination plays only a limited reproductive role. By contrast, in reflective judgment, the imagination is free to play *between* sensation and a kind of formal judgment. A reminder of Schiller's understanding of aesthetic judgment is helpful at this point:

> The mind travels from sensation to thought through a middle disposition in which sensuousness and reason are lively at the same time and for that very reason take away each other's determining force and bring about a negation by way of an opposition. . . . Thus one must call this condition [attunement] of real and active determinability of the aesthetic condition.[8]

While Schiller and Kant identify a certain reciprocity between sense and understanding in reflective judgment, a reciprocal process that may *seem* to be reflected in Tucker's discussion of bidirectional neural pathways, these thinkers are still operating under a kind of faculty psychology that situates the seats of reason and sensation in particular loci. For them, the imagination serves as the process by which these regions are mediated. Tucker's work dispels any lingering commitments to the faculty psychology model. His findings indicate that meaning, concepts, and memory are not developed under the auspices of the faculty of reason; rather, "reason" develops in the movement of activation traffic between visceral centers and more differentiated neural areas that map the sensory modalities. If this is the case, one might ask: "Where is the imagination in our description of thinking?"

The true insight in regard to the imagination—an insight that Kant and Schiller glimpse, an insight that Peirce begins to develop, an insight that Tucker unpacks to an impressive degree—is that the imagination is to be associated with a cognitive *process* laden with feeling and meaning. It is a process that at once reflects an activity and a genuine receptivity. This active receptivity develops over time in order to bring about a harmonious relation between a particular organism and the world. The process of the imagination is one that cannot be described exhaustively or determinately. It is embodied and develops spontaneously but, at the same time, is constrained and guided by past developments. It is a process that provides an effective bridge or interface between an organism, its embodied history, and novel environmental conditions. It is a process that "handles" the new possibilities that the world affords.

These points come to light in the details of Tucker's account. Depending on environmental circumstances, the neural traffic flows predominantly in one of two directions. Tucker explains that when memory or habit dominates a situation, the activation pattern radiates from the limbic system toward the sensory modalities. When novel sensations occur, however, the activation patterns reverse, proceeding from the somatic-sensory shell toward the visceral core, where it acquires meaning and associations. This is a temporal process in which new environmental stimulation continually affects, adjusts, and renews the structured activation patterns that form the physiological basis of human habit and conceptualization. Tucker develops an interesting metaphor when he writes, "it is like the mind breathes. In then out—weaving a confluence of distributed representations, weaving visceral meaning with external reality."[9] This process of weaving generates new patterns by returning to the latent patterns in our neural architecture and re-turning these patterns in novel ways.

It is worth noting that Tucker's physiological description supports the phenomenological accounts that John Dewey, following in the steps of Peirce's "Thinking as Cerebration," develops in his 1896 essay "The Reflex Arc in Human Physiology." Here, Dewey describes a child's process of grasping a hot candle, being burned, and subsequently learning not to grasp it. Dewey speaks at some length concerning the confluence of emotion, sensation, motor functions, and understanding:

> Let us take for our example, the familiar child-candle instance.[10] The ordinary interpretation would say the sensation of light is a stimulus to the grasping as a response, the burn resulting is a stimulus to withdrawing the hand as response and so on. [But] upon analysis, we find that we begin not with a sensory stimulus, but with a sensori-motor coordination, the optical-ocular, and that in a certain sense it is the movement which is primary, and the sensation which is secondary, the movement of body, head and eye muscles determining the quality of what is experienced. In other words, the real beginning is with the act of seeing; it is looking, and not a sensation of light. The sensory quale gives the value of the act, just as the movement furnishes its mechanism and control, but both sensation and movement lie inside, not outside the act. Now if this act, the seeing stimulates another act, the reaching, it is because both of these acts fall within a larger coordination; because seeing and grasping have been so often bound together to reinforce each other, to help each other out, that each may be considered practically a subordinate member of a bigger coordination. More specifically, the ability of the hand to do its work will depend, either directly or indirectly, upon its control, as well as its stimulation, by the act of vision. If the sight did not inhibit as well as excite the reaching, the latter would be purely indeterminate, it would be for anything or nothing, not for the particular object seen. The reaching, in turn, must both stimulate and control the seeing. The eye must be kept upon the candle if the arm is to do its work; let it wander and the arm takes up another task. In other words, we now have an enlarged and transformed coordination; the act is seeing no less than before, but it is now seeing-for-reaching purposes. There is still a sensori-motor circuit, one with more content or value, not a substitution of a motor response for a sensory stimulus.

Dewey continues by explaining how these initial stages of sensory coordination bleed over into the process of learning and remembering. He emphasizes how learning helps determine the course of future bodily coordination:

> Now take the affairs at its next stage, that in which the child gets burned. It is hardly necessary to point out again that this is also a sensori-motor coordination and not a mere sensation. It is worth-

while, however, to note especially the fact that it is simply the com-
pletion, or fulfillment, of the previous eye-arm-hand coordination
and not an entirely new occurrence. Only because the heat-pain quale
enters into the same circuit of experience with the optical-ocular and
muscular quales, does the child learn from the experience and get
the ability to avoid the experience in the future.[11]

Dewey did not have the resources of contemporary cognitive neuroscience, but his intuitions in regard to human learning have been borne out
by recent empirical studies. As Tucker underlines, the associations that
are made in the process of human learning depend on (1) the simultaneous activation of multiple sensory modalities, what Dewey calls the
"seeing" of the candle; (2) the spreading of activation patterns along
corticolimbic pathways toward the visceral core; (3) the structural adaptation of the limbic core that provides, in Tucker's words, the "motive
direction of memory integration" (Dewey roughly echoes this point
when he remarks that the burned child only learns and remembers when
the "heat-pain quale enters into the same circuit of experience with the
optical-ocular and muscular quales"); and (4) an integration of sensory-
motor responses that depends on the visceral core's ability to associate
particular modalities—such as seeing and grasping—over a period of
time. Dewey begins to touch on this phenomenon, stating that "seeing
and grasping have been so often bound together to reinforce each other,
to help each other out, that each may be considered practically a subordinate member of a bigger coordination." This bigger coordination is
embodied by deeper neural activation patterns established by the repeated reinforcement of particular stimuli.

Neural Reentry and Imaginative Coordination

We have begun to describe the embodied character of the imagination.
We have also seen that the emergent and adaptive disposition of the
imagination arises from the physical/neurological processes of a physical organism. It seems necessary to describe more fully the neural processes by which various environmental stimuli (input) establish neural
patterns but also give rise to creative variations. In effect, we turn our attention back to the neurons that constitute significant parts of the brain

and ask how their aggregation and selection structure the relations of an adaptive mind. In Tucker's work, it is sometimes easy to forget that the "shell" of the brain is not simply a self-contained entity but rather constituted by relations of an estimated ten billion interconnected neurons. Each neuron receives synaptic input from many thousands of other neurons such that within each cubic millimeter of the brain's gray matter there are approximately one billion synapses. We must think through this structural complexity to think through the imagination.

I will focus on the concept of neural reentry and degeneracy in order to examine the aggregation and selection of complex neuronal populations.[12] These concepts lead to an investigation of complexity in a broader biological setting. Here we ought to regard a complex system as "one in which smaller parts are functionally segregated or differentiated across a diversity of functions but also as one that shows increasing degrees of integration when more and more of its parts interact."[13] Reentry and degeneracy begin to explain the way in which neural dynamics are characterized by this differentiation and integration. More distantly, they point to the process known as *autopoeisis* and complex agency, which will be addressed in the coming sections. Any description of reentry and degeneracy should be regarded as another articulation of cognition that seeks— once again—to overcome the legacy of the Cartesian divide between matter and mind. In this respect, this rendering is another attempt to deepen an understanding of the imaginative processes described earlier. In all of these discussions, we should listen for echoes of Peirce and even Kant; their insights into the nature of the imagination resonate closely with these contemporary empirical accounts.

In the previous sections, we addressed the correlation of selective events across various maps of the brain. This process has been described as being driven by the mechanism of neural reentry, a concept that rests at the heart of Edelman's research. Recognizing that the primary consciousness of human beings is characterized by an integrated *Gestalt* that continually adjusts to one's surroundings, Edelman describes a process of neural integration that might begin to ground this conceptual continuity.

As the work of Sporns, Edelman, and Tononi indicates, reentry may at first glance resemble a kind of biological feedback between brain regions, but it differs from feedback in several important respects. Feedback (here it might be helpful to think of the workings of a mechanical thermostat) operates along, and as, a single fixed loop made of reciprocal connections using previous instructionally derived information of control and correction (known as an error parameter). In contrast, reentry is a *selectional* and *distributed* rather than an *instructional* system, which co-adapts over time *without* a specific and predetermined error parameter. Reentry occurs across *multiple* parallel pathways, connecting multiple synaptic maps, and provides for the co-evolution of these maps and connections over time.[14]

The most important distinction to be made between feedback and reentry is the fact that reentry has a constructive and reconstructive function rather than the merely corrective one demonstrated in feedback. Reentry is constructive in the sense that it coordinates functionally separate neural maps in developing new neural activation patterns and in reestablishing and refining preexisting patterns. Reentry appears to reflect the evolutionary and imaginative dynamics that proceed from past forms while extending them in novel ways. That is also to say that reentry is directed and irreversible.

We touch on an issue that Peirce addresses in his work on continuity and generality. As mentioned earlier, in the "Law of Mind" Peirce states that "logical analysis applied to mental phenomena shows that there is but one law of mind, namely that ideas tend to spread continuously and to affect certain other ideas which stand to them in a particular relation of affectability."[15] With reentry, we begin to explore the physiological basis of an idea's tendency to "spread continuously" over time. The concept of reentry trades on a question that Peirce asks in the 1890s: "What can it mean to say that ideas wholly past are thought of at all any longer? . . . How can a past idea be present?" Researchers are beginning to answer Peirce by identifying physical recurrent processes that underpin the continuity of human thought. The neural process that Edelman outlines in reentry is a recursive one that mediates between past patterns of activation and adaptive structural developments that respond to the novel and problematic aspects of environmental conditions as they arise. Edelman suggests that reentry is the neurophysiological foundation of the "remembered present" that defines human consciousness. The recursive activation of neuronal systems allows organisms that act in the moment, and in a particular problematic situation, to redeploy patterns of behavior that remain continuous with past forms. In terms of reentry's function in the development of human conception, it *reconstructs*, literally "piles together again," the various qualities and aspects of our perceptual fields in the coherence of primary consciousness. This coherence, however, does not preclude the possibility of novel forms of categorization and coordinated motor responses that arise in light of and seek to respond to surprising environmental conditions.

The process of neural reentry might be understood more easily through an analogy to the development of improvisation in a group of musical artists. Imagine a jazz quartet in which each player responds to ongoing cues of her own playing but also the cues and tempo of her accompanying players. No sheet music is used, and in the opening moments of the first movement the styles and tempos are organized around a general theme. Over time, a more specific beat and theme are established as the musicians begin to correlate, or "get in time." If the musicians have been playing together for many years, certain musical signals seem instantaneously to connect the four musicians, causing a deepening of the correlation and resonance of sound. Each new signal causes a wave of novel sounds that, in the midst of novelty, maintains harmony with past forms. As Edelman and Tononi conclude from a similar analogy: "Such integration would lead to a kind of mutually coherent music that each one acting alone could not produce."[16] They elaborate on this metaphor, stating that just as two pieces of music are never identical, the dynamics, relations, and connections of two brains are never exactly alike. Reentry begins to point to a mechanism that might enable this integration. In addition to being crucial to signal integration, "specific linkages of a reentrant type between two sets of groups can lead to the emergence of new associative functions not originally in either set of groups."[17]

This analogy between reentry and music is instructive in two significant respects. First, it provides a way of understanding the synchronization of reentry—the reciprocal and parallel signaling that underpin the process. More significantly, this particular analogy of musical improvisation sheds light on the nature of the imagination. Musical improvisation, the beautiful give and take of embodied artists, has just been described in brief. Novelty and continuity are its defining marks. Underlying this aesthetic improvisation is embodiment—the physical processes of the musicians that grant the possibility of imaginative cognition. These processes themselves, including the process of neural reentry, help explain the reciprocal *Spiel* that has long defined the realm of aesthetic judgment and production. Indeed, these neural-biological processes are best described by way of analogy to the phenomenological accounts of artistic creation. The imagination is seen once again not as being a peculiar aspect of human aesthetics but rather as an aspect of the very organization through which conscious life arises.[18]

Edelman elaborates on the concept of neural reentry, noting that it depends upon the anatomical precondition of "the remarkable massively parallel reciprocal connection" of the brain areas.[19] This parallel and reciprocal—and one might add "mediating"—character of brain processes is exemplified by the corpus callosum, a huge bundle of reciprocal fibers connecting the two cortical hemispheres. This structural bridging, most dramatically demonstrated in the corpus callosum, takes on a variety of complex forms that constitute the structure and function of the brain; other parts of the brain are coordinated by similar means. In the case of the corpus callosum, the right and left hemisphere synchronize in an ongoing, recursive interchange. Reentry depends on and is defined by this reciprocal interchange of parallel signals between connected areas of the brain that coordinates the activities of these area's maps. As Edelman notes: "The most obvious abnormality in people with split brains . . . is a profound deficit in the interhemispheric integration of visual and motor information. These persons are not able to integrate visual information presented to their two visual half fields."[20] In this sort of disorder, the reentrant mappings that usually serve to connect functionally distinct brain regions reciprocally are severed. This physiological severing causes a person's conceptual space to be torn asunder in particular ways. Our discussion of degeneracy in the coming section will readdress this situation and explain why such injuries do not cause global cognitive dysfunction.

As Edelman suggests, reentry and its anatomical substrate are crucial in the development of conceptual integration, which in turn is essential for the creation of a cohesive scene in primary consciousness. "Integration can best be illustrated," Edelman writes, "by considering exactly how functionally segregated maps in the cerebral cortex may operate coherently together even though there is no superordinate map or logically determined program."[21] If the experience of making music, gardening, or any other creative event is multifaceted, holistic, and continuous, there must be a corresponding neural integration that serves the condition for the possibility of this experience.

Degeneracy and Neural Plasticity

In describing the mechanism of reentry it is also necessary to address the neural property of degeneracy, mentioned in passing earlier, which seems to demonstrate the imaginative and probabilistic nature of synaptic operations and points to the physiological basis of functional adaptation. Perhaps more distantly, the concept of degeneracy may help explain the notion of conceptual flexibility and metaphor. Research on degeneracy has shed more light on the plasticity and complexity that defines neuronal systems. In common parlance, the term degeneracy reflects a pejorative connotation of being deficient or degraded. Researchers in general biology and cognitive neuroscience employ the term somewhat differently, drawing on its meaning in differential calculus. In this context, degenerate equations are those that possess a common solution. As Edelman and Tononi explain in their application of degeneracy in studies of biology, "degeneracy is reflected in the capacity of structurally different components to yield similar outputs or results."[22]

Degeneracy should not be confused with the redundancy that occurs when structurally identical elements produce the same result. Edelman and Gally note this point in their remarks concerning the difference between creation by design and creation by selection:

> The contrast between degeneracy and redundancy at the structural level is sharpened by comparing design and selection in engineering and evolution, respectively. In engineering systems, logic prevails, and, for fail-safe operation, redundancy is built into design. This is not the case for biological systems. . . . In general, an engineer assumes that interacting components should be as simple as possible, that there are no unnecessary or irrelevant interactions, that there is

an explicit assignment of function . . . to each part of a working mechanism, and that error correction is met by feedback. . . . By contrast, in evolutionary systems, where there is no design, the term "irrelevant" has no a priori meaning. It is possible for any change in a part to contribute to overall function.[23]

They elaborate on these points by claiming that, unlike the structures of engineering, the structures of evolution are not assigned *exclusive* responsibility for a particular function. This is also to say that evolutionary systems are defined not necessarily by simplicity but by the propagation of the complex interactions of their parts. The flexibility and compensatory effects of *degeneracy* are seen in many organic systems. For example, many different DNA sequences can specify the same amino acid. Different subsets of genes can cause the same phenotypic structure. Different antibodies can identify and counteract the same foreign body.

Because of the immense complexity of neuronal populations of the human nervous system—cited in the section on reentry—its degree of degeneracy is far more extreme than forms examined in other cellular or genetic systems. Despite the *extremely* large number of neurons in most vertebrate nervous systems, no two neurons are exactly alike. Even in genetically identical organisms, no two neurons are morphologically identical. This being said, morphologically different structures can perform the same function, just as many structurally dissimilar keys can open the same door.

Think about an apartment complex with many different apartments and many different tenants. *Each* tenant needs a key that opens her apartment and her apartment only. *Each* key would be structurally unique to serve this particular function. Additionally, however, *every* tenant would also need access to the laundry room and the room in which the trash and recycle cans are kept. Instead of producing redundant keys that opened these two rooms and distributing them to the tenants, the apartment manager makes a decision to produce a degenerate key system in which the apartment keys (remember that these must be structurally different) also open the two public rooms. Hence, different structures perform the same function. Now let us consider the advantages of this degenerate system.

The first advantage has to do with the *durability* of the degenerate system. Let us think about the laundry room key example. If there were only one key that could open the laundry room door, meaning if there was a one-to-one correlation between the morphology of the key and the function of opening a particular lock, this would be a fragile system indeed. If this one key was lost, destroyed, or altered in the course of mutation, the function would be cut out of the repertoire of activities the system could undertake. Nobody would get their laundry finished. Now, imagine if getting one's laundry finished was a vital function for the system (perhaps for some of us it truly is). If this were the case, a single mutation (say, a chipping or bending) or misplacement of the key would have smelly, disastrous, or even lethal consequences. Let us say that we want to avoid this situation. We could go about it in two distinct ways. First, we could create redundant keys that opened the same lock. Very good. Now all of us can open the door, and if one key is lost, damaged, or altered by mutation we need only find another copy of the key. There is, however, a cost to redundancy: resources have to be allocated for the construction of these multiple copies. There is a solution that might be more viable and would be favored in terms of evolutionary advantage—the degenerate solution. In this case, the keys would be made so as to be morphologically dissimilar yet have the same basic function of opening the public room. One might wonder how this solution would have any advantage over a redundant one.

The answer, I believe, brings us to the second advantage of degeneracy, the advantage of the *flexibility* and *novel adaptation* that degeneracy affords. In a degenerate system, the morphological dissimilarity or diversity has an interesting value, a value *in posse* and *in actu*. In light of new environmental conditions, these diverse forms can be put to more specific uses, often with dramatically positive consequences. For example, my key will allow me to open the public room, but thanks to its higher degree of structural differentiation (the extra little notch on the end of the key) I am also able to perform another function, namely, opening my own apartment. This is an advantage that redundancy could never claim. The surplus of differentiation and diversity has a *potential* advantage to a system. It is worth noting that adaptive systems, such as the ones that operate under the rules of natural selection or neural

Darwinism, could both generate this diversity and benefit from the degenerate properties that such complexity yields. On the other hand, degenerate flexibility is notably missing in the instructional repertoire of most artificial computational systems.

While the laundry room key example might be helpful for certain purposes, it ought to be accompanied by two caveats. First, I would like to take a closer look at the structural character of keys. We usually regard a key as a *single* thing that hangs on a key chain and is used for a certain purpose. The ability of a single key to function properly, however, depends on the *particular configuration of its constitutive parts* that make up its ridges and valleys. An infinite but determinable variety of configurations *could* possibly produce the same output or resultant function. That is not to say, however, that *any* configuration *will* produce the same function.

The second cautionary word stems from the first. If I ask you: "What opens that lock?" You could simply and easily say: "That key." It is normal to think of one key opening one lock. Degeneracy, however, problematizes this question by suggesting that many different keys or pathways could open the self-same lock—not to mention all of the hairpins, skeleton keys, credit cards, and pick axes that could "unlock" the laundry room door.

On the synaptic level, degeneracy is seen when a particular environmental stimulus (or firing from another part of the brain) causes *any* set of neural circuits to fire in a set of output responses *similar* to those that were previously adaptive and thus provides the basis for repeated mental or physical acts. More simply, a *similar* circuitry response can re-create specific mental or physical acts over time. Similar output responses refer to a relational similarity in which the relations between particular neural groups tend to be roughly or approximately the same. Edelman elaborates: "So what is called forth when an act is repeated must be any one or more of the neural response patterns adequate to that performance, not some singular sequence or specific detail."[24] This is a critical point. He explains that when it comes to selective cognitive systems, "there is no unique structure or combination of structures corresponding to a given category or pattern of output. Instead more than one combination of neuronal groups can participate in more than one signaling function."[25]

Degeneracy may help explain how, in an ever-changing environment, the embodied mind has the ability to develop a stable, albeit flexible, pattern of behaviors and categorizations.

Peirce's speculations concerning generality and continuity anticipate Edelman's emphasis on the flexibility that characterizes neural degeneracy. For Peirce, generality, continuity, and vagueness are all, in the words of Carl Hausman, "species of indeterminacy." Anticipating complexity theorists, the early pragmatist insists that "diversification and specification have been continually taking place" in every natural and physical system.[26] For this reason, when you try "to verify any law of nature, you will find that the more precise your observations, the more certain they will be to show irregular departures from the law."[27] This suggestion is borne out directly in the empirical investigations of degenerate neural dynamics.

At first glance, the recategorization and generalization that coincide with reentry and degeneracy, however, seem to have particular limitations. How do genuinely new concepts emerge—as they do in a child's cognitive development or in the genius's artistic development? Peirce's comments on irregularity and indeterminacy help explain this novelty. In effect, we are revisiting the relation between synechism, characterized by continuity, and tychism, characterized by spontaneity. Edelman responds to this problem in somewhat vague terms. He states that "when changes in synaptic efficacy occur in neural systems . . . they allow the possibility of further refinement or alteration of perceptual categorization."[28] He attempts to address this question in more concrete terms when he turns his attention to the complexity of the reentrant interactions of neuronal groups. Neuronal groups are connected to a vast number of other groups regardless of spatial proximity. These many groups are reciprocally connected and functionally distinct from the rest of the brain. Because of this connectivity, small changes in environmental situations can cause new conceptual associations to be made and new behaviors to arise since "any subtle change in activity of different regions of the brain can bring about new, dynamic associations."[29] Depending on the consequences of these associations, depending on the way in which these associations might enable an organism to negotiate its ever-changing environment, these novel constructions may be reinforced and "learned" as they impress themselves on the dynamics of neural organization. Creative novelty does not depend on a certain supernatural élan vital that animates matter but rather on the massively complex interactions between natural agents that grant the possibility of nonlinear effects.

In discussing reentry and degeneracy, one is at least indirectly discussing the character of human memory. While a detailed treatment of this point cannot be gone into here, it does seem to be a worthwhile prompt for future investigation. Degenerate and reentrant neural circuits allow for changes in memories and concepts as new experiences occur and a new environmental context evolves. Memory in a degenerate-reentrant system is "recategorical," or, speaking in the now-familiar language of the imagination, re-creative rather than strictly repetitive.

Recognizing Complexity: Imaginative Synthesis and Conceptual Forms

The way in which degenerate neural circuits operate by way of correlation and relative similarity and difference has been clarified in Peter Gardenfors's investigations of conceptual space in the field of cognitive neuroscience. Here, we are briefly shifting levels of analysis. To avoid confusion, it is necessary to recognize that Gardenfors is not approaching the question of cognition from a biological vantage point but from the perspective of computer science. While this distinction should be kept in mind, it is also necessary to draw out the parallels between his understanding of the formation of concepts and the aforementioned discussion of neural activation and aggregation. In the description of consciousness that is underway, we are attempting to describe multiple co-emergent systems in which imaginative processes seem to emerge. An investigation on several seemingly disparate levels of analysis may, at first glance, seem confusing, but it is necessary in order to describe the nesting of imaginative systems. Gardenfors's work leads us deeper into the nature of the imagination while reminding us that neuronal dynamics may prove isomorphic with the human formation of concepts. One cautionary word regarding Gardenfors's work: in his development of parsimonious models, Gardenfors occasionally sacrifices the complexity of biological systems in order to amplify the explanatory power of his computational simulations.

Gardenfors extends the 1980 studies of Chipman and Shepard[30] and Edelman's recent work when he states that conceptual formation and integration depend on a mechanism that can detect and respond to similarity. Additionally, his research provides a computational approach that fits nicely with work on metaphor, prototype effects, and category structure. In terms of conceptual representation, Gardenfors writes, "representation is representation of similarities." At first glance, this comment does not seem to echo Lakoff and Johnson's understanding of metaphor. But it does resonate more closely as Gardenfors explains his position. He elaborates on his statement, insisting that "representations need not be similar to the objects they represent" but rather that "the representations preserve the similarity *relations between* the objects they represent."[31] In this light, it might be more accurate to say that representation is a type of correlation. This rendering of conceptualization stands as a radical departure from earlier models that sought to advance the idea that mental concepts held a one-to-one relation with the form of an external object.

Gardenfors also echoes comments made earlier with regard to conceptual prototypes by underlining the way in which conceptual formation reflects a degree of ambiguity that, far from being maladaptive, grants the possibility of mental flexibility and novelty. His work, like that of Lakoff and Johnson, serves as a reminder that the processes of neural reentry and degeneracy—processes that reflect a degree of spontaneity and flexibility—have a linguistic correlate. It points to the way in which the unique growth and creativity of biological networks continually provide the groundwork for the development of conceptual thought and language.[32] Gardenfors takes his cue from prototype theory but seeks to extend this work in important ways. He observes that "while prototype theory fares much better than Aristotelian theory of necessary and sufficient conditions in explaining how people use concepts, the theory does not explain how such prototype effects can arise as a result of learning to use concepts."[33] While Lakoff seeks to extend Rosch's understanding of prototype effects in his work on radial category structure, Gardenfors elaborates on Lakoff's understanding of category structure by attempting to model the process by which prototypes are established by exposure to experiential cases. Gardenfors employs computer science and particular computational models in order to describe the dynamic emergence of category structures over time.

Like Kant's understanding of the development of aesthetic ideas and Peirce's understanding of human cognition, Gardenfors believes that "we are not born with concepts, but they must be learned."[34] He elaborates by stating that, "the decision procedures that connect perceptions and actions must be created with the aid of the experience of an agent."[35] He reflects a type of pragmatic sensibility in his belief that concepts must serve as guides for action and must be judged upon their consequences and on their ability to interpret novel situations and patterns of experience. In his words, "to be useful, the procedures should not only be applicable to known cases, but should generalize to new situations as well."[36]

Gardenfors's model is relational (connectionist) and functional in the sense that cognitive relations change over time in order to accommodate new organism-environment relations. Gardenfors's work attempts to quantify the way in which radial category structure and conceptual prototypes are established and modified by experience. By employing vector geometry, he attempts to develop computational models that might accommodate and describe the transformation of conceptual prototypes. Earlier, we briefly discussed the way in which conceptual examples that emerge in the midst of experience establish conceptual norms, or prototypes. These are the prototypes or "typical instances" to which future cases are compared in accordance with the qualities and correlations that define this norm. Gardenfors proposes that researchers might employ Voronoi tessellations to model the dynamics of this process. The challenge for future research will be to explore the physiological correlates to this theoretical model.

Gardenfors suggests that we return to the idea that concepts are established over time by an agent's exposure to exemplars. These exemplars establish a prototype that then serves as a benchmark for the future experience of exemplars. He echoes Langley when he suggests that these prototypes (p) can be arrayed in two-dimensional conceptual or metric space as the mean (Σ) to the ith coordinate for all exemplars where x_{ik} stands for the location of n, such that the equation for this prototype (p) could be written: $p_i = \Sigma_k x_{ik}/n$. These prototypes then generate a

Voronoi categorization that can represent a conceptual domain. These domains would have convex borders that are established by equidistant points between said prototypes. Within these convex regions it is possible to say that members of the category are more or less central. Now, without considering a detailed schematic, it is possible to imagine the effects that a new exemplar that is experienced and placed at the edges of the Voronoi categorization might have on the position of a prototype and its corresponding categorization. Since the prototype is established as the mean of the placement of past experiential cases, the change of the mean will be a dynamic process that is tethered to the past actualities and accommodates new events. This dynamic process can be calculated for the effect "change in p" (Δp_i) on dimension i of the stored prototype p in the manner described by Langley in 1996:

$$\Delta p_i = (x_i - p_i)/(n+1)$$

where x is the ith coordinate of the newly learned instance and n+1 is the total number of instances (including the new one) in the class represented by the prototype p. Here, we can notice the similarity between this function and Hebbs rule.

Voronoi tessellations are often used to describe supervised learning, in which the learner is given feedback in the form of names or other standards of the error parameter in order to identify which exemplar belongs to which category. As Gardenfors notes, however, these functions can also model learning mechanisms that can "pick up patterns in the environment without any error messages concerning the performance of the mechanism. All such mechanisms need is a large sample of training examples."[37] These systems are called unsupervised learning modules, and they more closely approximate the creative learning that is distinctive in human cognition and neural development.

Unsupervised learning techniques have been a hot topic in computational research for more than thirty years and have recently converged with the computational studies of abductive processes. Gardenfors notices that these techniques are often reflected in clustering models that sort a stream of objects into various concepts by degrees of similarity and employ Voronoi tesselations as their computational basis. While such a prototyping task exemplifies the adaptability of these functions, it seems artificial in the sense that programmers, prior to the initiation of the task, must input certain lists of atomic features or attribute markers that serve as criteria that define the similarity measure. These features and attribute markers substitute, rather imperfectly, for the flexible and inherited modes of embodiment (nascent neural foundations, genetic architectures, and basic chemical relations) that ground, limit, and enable our thought-full lives. Instead of inheriting these cognitive constraints, programmers input the parameters of thinking.

This understanding of dynamic concept formation, formation that operates by way of similarity and difference, is interesting to the extent that it does not necessarily presume a one-to-one correlation between the objects of the world and our concepts of them. More importantly, perhaps, is the fact that the approximations and "fuzzy" categories allow us to make provisional and speculative inferences, the sort of inferences that occupied so much of our discussion of abduction and that have been readdressed in the previous section on reentry and degeneracy. These comments ought to remind us of the categorization that occurs in Peircean abduction. Peirce writes:

> Abduction furnishes the reasoner with the problematic theory which induction seeks to verify. Upon finding himself confronted with a phenomenon unlike what he would have expected under the circumstances, he looks over its features and notices some remarkable character of relation among them, which he at once recognizes as being characteristic of some conception with which his mind is already stored, so that a theory is suggested which might explain that which is surprising.[38]

Flexible categories, instantiated or, in Peirce's words, "stored" in our neural architecture and dynamics, in effect form the basis of our flexible hypotheses, those tentative "theories" that might explain what is surprising in our experience. The consequences of these hypotheses—calculated as a function of fitness of the neural system and, by proxy, the organism on the whole—provide a selection mechanism by which future similarity measures are established.

The complexity theorist John Holland writes that "to know an instance of a member of a natural category is to have an entry point into an elaborate default hierarchy that provides a wealth of expectations about the instance."[39] This default hierarchy, stored in particular patterns of neural activation, is presumably established over a period of time in which an agent is exposed to a large sample of training examples. In a naturalized setting, these training examples would be experienced over a period of time and would be associated with a series of fitness consequences that would be experienced as being emotionally meaningful. These consequences would slowly create the "wealth of expectations" about future instances. In this case, it is worth noting that specific expectancies are slow in development. Children typically overgeneralize a category and face the rather rude consequences of their categorical generalization. For example, in a naturalized setting the concept "berry" might be associated with the consequence of getting something sweet to eat. From the experienced fact that some, or even most, berries are sweet, the child generalizes to conclude that all berries are sweet. In this case, the particular light green berry that the child spots would be plucked with little hands and shoved into a little mouth. Upon chewing this light green berry, however, another sort of consequence might ensue: this light green berry tastes horrible and makes the toddler sick. We are reminded of Dewey's discussion of the reflex arc and notice that the neural patterns that led to categorical generalization have been interrupted and altered by the unpleasant consequence that cuts in on our habitual experience.

To say that these neural patterns are interrupted and altered is also to say that our physiology is irrevocably revised in order to prepare us better for future encounters of a similar kind. Granted, these revisions are usually small—perhaps too small to equip us with the hypotheses that might be required for the coming experiences. Remember that changes in the Voronoi tessellations occur by the calculation of the *mean* of the sample of experiential cases. Metabolic dynamics described by Hebb must be calculated over an entire population of neural aggregates that serve as a sort of ballast for any one particular and novel activation pattern. Over enough time, however, exposure to new exemplars may either shift the meaning of a concept, enlarge the concept (demonstrated by generalized Voronoi tessellations), or establish a new concept with more differentiated hypotheses.

For example, upon observing that this object is like a berry, I might first hypothesize that this object is edible. Upon observing that this object is like a mulberry or juniper berry, I would no longer make this hypothesis. The distinction between a berry, in general, and a mulberry in particular is developed over time and sets the stage for novel hypotheses generated on the ground of similarity of difference.

The picture of the brain that emerges from these hypotheses portrays a dynamic system that is highly integrated but also highly differentiated. It is a system where constituent parts coordinate and adapt despite—indeed, because of—external and internal stimuli. This quality of being simultaneously differentiated and integrated has already been used to describe the imaginative unity of apperception in Kant's first *Critique*. It is also reminiscent of the way in which the imagination provides the possibility of aesthetic novelty while at the same time making this novelty in some sense continuous with past forms. Remember that the imaginative genius, for Kant, serves as the bridge between the historical canon of art and novel forms of aesthetic production. It has also been used by Kant to describe the way in which organisms develop over time. Similarly, this quality has been used to describe Peirce's notion of continuity, which seeks to describe the vitality of an organic *Gestalt* while still emphasizing the differentiation of its particular parts. In this sense, the "human mind" ought to be regarded as the relational processes of a complex system, a system of nested systems that orchestrate and co-evolve over time. It must be made clear that the complexity of the mind cannot be fully understood as a particular "synaptic system," as the "limbic system," as the "cortical systems," or as the organic system of the human body but rather as a complex system that includes, as nested relations, all of these networks.[40]

Autopoiesis and Agency

This very brief description of neural networks begins to point to general principles that may help us ground a future discussion of an imaginative ontology. A few guidelines should be kept in mind. Neural networks reflect three distinct qualities: the system is self-bounded, self-generating, and self-producing. To be *self-bounded* means that the system's extension is determined by a boundary that is an integral part of the network. Notice that this definition does not preclude the possibility of growth.

To be *self-generating* means that all components, including those of the boundary, are produced by processes within the network. Notice that this definition does not preclude the possibility of interaction between autonomous networks. To be *self-producing* means that the production processes continue over time, so that all components are continually reinstantiated, renewed, and possibly replaced by the system's processes of transformation. Notice that this definition does not preclude, and indeed insists upon, the possibility of novel organizational structures. Our investigation has once again led us to three characteristics that echo the earlier findings in the discussion of the imagination. These three characteristics emerge in philosophic form in Kant and Peirce but appear again in recent empirical studies in neuroscience and complexity theory. These three characteristics are the lynchpins of any *autopoetic* system, as described by Humberto Maturana and Francisco Varela.[41] Maturana and Varela hypothesize that *autopoetic* characteristics define not only the material life of human consciousness (the development and growth of neural networks) but *any* form of life.

Let us return to the short vignette that opened this project, to an instruction that was as confusing as it was emphatic: the invitation to *"Be imaginative!"* In the first four chapters of this book I have tried to map the movement of the imagination in the work of Kant and Peirce. In doing so, I described the way in which their thinking through the imagination leads them from a relatively narrow conception of imagination as an aesthetic or epistemological concept to the development of a broader rendering in which imaginative action and ordering are regarded as the keystones of their respective epistemological models.

Imagination takes root in and arises from the processes of natural being and seems to reflect Parmenides' comment referenced in chapter 4 that "the same thing is for thinking as for being." A reexamination of this comment will serve as a conclusion and an opening for further research. The work in empirical psychology and cognitive neuroscience leads us to reconceive the character of natural being and develop a particular ontology that can account for the shared imaginative structuring of human conceptualization and natural systems. According to the neurobiologist Vittorio Gallese, this reconceiving is itself, very literally, natural:

The brain builds an ontology, an internal model of reality, which—on a very fundamental level within its representational architecture—incorporates the relational character between inter-actions between organism and environment, and that this architecture can actually be traced at the microfunctional level implemented in the brain's neural networks. The same subpersonal ontology then guides organisms when they are epistemic agents in a social world: Interpersonal relations become meaningful in virtue of a *shared action ontology*.[42]

Gallese's observation is helpful in the sense that he, in the spirit of Peirce and other American philosophers including Josiah Royce, William James, and Alfred North Whitehead, identifies a particular type of relational process as the building block of being. It also suggests, as Metzinger states, "phenomenal experience of being an intentional agent, of being a perceiving, attending, and cognizing subject can be naturalized."[43] Gallese, however, stops short of implying that this process of "naturalizing" human consciousness would involve attributing particular forms of agency and imaginative consciousness to nature itself. In the end, he seems to want no part of naturalizing agency. In this sense, he seems to diverge from Kant's rendering of purposive nature in the final sections of the *Critique of Judgment* and Peirce's comments on the "reasoning and the logic of *things*." Indeed, at one point Gallese states that "actions are elementary building blocks of reality for *certain* living organisms . . . bodily movements [as opposed to conscious actions] are simple physical events, and they can be represented accordingly." Likewise, behaviors, "as simple motor acts . . . also do not have a consciously experienced reward-producing component." In his hesitation, he distinguishes both movement and behavior from what he calls "proper action," seemingly reinstantiating a type of division between the material and the psychical.[44] This distinction can be analytically useful. But as pragmatists such as Peirce recognize, such distinctions have a way of becoming reified in unproductive ways.

Without succumbing to a mystical holism or naïve panpsychism, our discussion of the imagination seems to indicate that a bolder "shared action ontology" might be developed that accommodates the most recent empirical research while, at the same time, suggesting a new way to think about consciousness and agency. In the development of this process ontology, it seems appropriate to return briefly to later works of Kant. In the *Critique of Judgment*, in a project that centers on the processes and structure of the aesthetic imagination, Kant is led into a reevaluation of the natural world. We have followed Kant's train of thought from philosophy of mind to natural being in earlier sections of this book. By way of review and in brief, he argues that organisms, in contrast with machines, are self-reproducing, self-organizing wholes. In a machine, according to Kant, the parts only exist *for* one another, in the sense of supporting one another within a functional whole. In an organism, the parts also exist by *means* of one another, in the sense of producing one another. Already in the 1790s, Kant writes that "we must think of each part as an organ that produces the other parts so that each reciprocally produces the other . . . because of this [the organism] will be both organized and self-organized."[45] This comment resonates with points made earlier concerning the receptive-active disposition of the imagination. In saying that the organism is both organized and self-organized, Kant catches a glimpse into the complexity and adaptation of natural processes, into the ways in which the organic creativity that underpins our imaginative endeavors reflects a similar receptive and active character.

Peirce takes a similar ontological turn of thought in the late 1880s, especially in the *Monist* essays between 1887 and 1893, with his development of a conception of nature and mind in which nested networks of self-organizing agents co-evolve over time. Kant's comments on self-organizing systems, which stem directly from his work on the imagination, anticipate Peirce, whose work in ontology and metaphysics in turn foreshadows the very recent study of complex systems. I would point to the work of two complexity theorists, John Holland and Stuart Kauffman, to extend the discussion of the imagination and suggest a way of being imaginative.[46]

John Holland's *Hidden Order: How Adaptation Builds Complexity* attempts to expose the extent to which the natural world can be understood through models of what he calls "complex adaptive systems" (*cas*). These models are not only regularly used to describe the development of neural networks (feedback, reentry, degeneracy, and multimodal processes) but to describe the interaction of all complex "agents"—from antibodies in the immune system, to bacteria in a gut, to banking firms in Manhattan. Holland defines an agent by way of its behavior, stating that to be an agent, its behavior must be determined by a collection of rules.[47] This is not to say that agents cannot adapt over time but rather that behavior must be directed. Indeed, as these agents begin to interact and aggregate, the rules of the collective become surprisingly flexible and adaptive; neural degeneracy and reentry are instructive cases.

As seen in the evolution of neural networks, complex adaptive systems exhibit high degrees of aggregation in which less complex agents interact, producing complex large-scale behaviors. "Aggregates so formed," Holland writes, "can in turn act as agents at a higher level—meta-agents."[48] The behaviors of these meta-agents cannot be defined in any determinate fashion. Aggregation of individual agents produces nonlinear effects that cannot be predicted or generalized. Holland concludes that "perpetual novelty is the hallmark of *cas*."[49] Such new effects are termed emergent phenomena. A brief rendering of emergence will hopefully draw out the similarity between a crucial property of complex adaptive systems and characteristics revealed in earlier attempts to think through the nature of the imagination.

It will be important to note that the aggregation to which Holland refers does not sacrifice the individuality of its constitutive agents. Complex adaptive systems reflect aggregation but also high levels of diversity. In fact, emergent phenomena depend on the less complex systems' ability to maintain autonomy. In terms of *cas*, to act autonomously is *already* to act relationally. Echoing Peirce's understanding of tychism and evolution, Holland points out that "diversity is neither accidental nor random. The persistence of any individual agent, whether organism, neuron, or firm, depends on the context provided by the other agents."[50] The imaginative theme of freedom-within-constraint shows itself again quite suggestively. Not only is the negotiation of this odd tension the sole way of thinking through the imagination. It is *very literally* the sole way of making our way through natural life itself.

NINE

BE IMAGINATIVE! SUGGESTION AND IMPERATIVE

Surveying the Field

At the end of an afternoon of gardening, one looks back, if only for a moment, to survey the ground that one has covered and worked through. It seems fitting, therefore, to take account of the moves made in this book. I have argued that the imagination plays a central role in the development of human cognition and, more generally, in the human-ness of human life. At the service of this argument, I have amplified, extended, and explained the opening passage of *Art as Experience* that tells us: "the imagination is a way of seeing and feeling things as they compose an integral whole. It is the large and general blending of inter-ests at the point where the mind comes in contact with the world. When old and familiar things are made new in experience there is imagina-tion."[1] In my examination of the role of the imagination in the develop-ment of German idealism, American pragmatism, and the contemporary cognitive neurosciences, I have identified three movements of the imagi-nation that anticipate or echo this quotation.

First, I addressed the critical works of Immanuel Kant, concentrating on the way in which the concept of the imagination is developed in the *Critique of Pure Reason* and the *Critique of Judgment*. I provided a reading of Kant's rendering of aesthetic judgment, common sense, and genius that rescues the productive imagination from its narrow artistic interpretation and restores it as a vital aspect and pervasive force in human cognition. In his account of aesthetic judgment, we see Kant thinking through a dynamic process in which the things of the world are seen and felt as a unified whole without appealing to the determinate rules of the understanding. It is important to note that this seeing and feeling is not constrained to the perception of an isolated individual but rather is communicable by way of a common sense. In the third *Critique*, Kant suggests that the seeing and feeling of aesthetic judgment is possible only if our minds are in contact with the world. This point is made explicit in his discussion of genius, in which he claims that genius is a "*natural* gift," a gift by which nature gives the rule to art. This rule places some natural constraints on the creativity of genius but also enables its generative force. Here, Kant seems to reflect the pragmatic sensibility that the world is not "out there"—something alien to and apart from human inquiry. Instead judgment and inquiry have access to the world only because they are always embodied in its movement. He acknowledges the movement and novelty of thought as he develops the sections on genius, reframing the imagination to emphasize its productive rather than its merely reproductive role. He initiates this project in the *Critique of Pure Reason* but extends it in detail in his description of aesthetics. In so doing, he sets the groundwork for a belief that when the old is made anew in experience, there is imagination.

Although Kant begins to think through the imagination, he leaves the project largely unfinished, failing to explain how creative imagination and reflective judgment might apply to cognition on the whole. The work of C. S. Peirce appropriates Kant's understanding of aesthetic judgment (via Schiller) and develops this project with his initiation of the pragmatic movement in the late 1860s. Like aesthetic judgment, pragmatic inquiry does not operate by way of determinate and instructional rules. Rather, it proceeds by way of a certain sensitivity to a type of harmony that might be established between an agent of inquiry and wider environmental situations. That is also to say that pragmatic investigation proceeds imaginatively by pursuing and remaining sensitive to a feeling of a unified whole. For Peirce the unity that is reflected in human conception can be traced to the dynamics of the imagination. Peirce's epistemological position rests on the insistence that human knowledge is always "in touch" with the natural world, indeed, that the mind is always continuous with the physical processes of human comportment. This continuity is reflected in the 1890 *Monist* articles but also poignantly expressed in earlier pieces such as "Thinking as Cerebration."

One of Peirce's key insights is that human thinking is defined by continuity but also by its creative spontaneity. As Kant notes, the imagination trades on the maintenance of continuity but also the production of novelty. American pragmatism takes up the issue of novelty by turning its attention to the character of hypothesis formation and the way it furnishes unforeseen answers to questions that continually confront us in experience. I have highlighted Peirce's development of abduction, musement, tychism, and synechism as important moments in which imagination receives the attention it is due. In describing tychism and synechism, Peirce makes a radical claim, namely, that the novelty that characterizes human imagination is rooted in the spontaneous processes of the natural world. These are physical processes that cannot be described by mechanical rules or by probabilistic frequency.

While Peirce's 1898 claim that "things live and move and have their BEING in a logic of events" remains highly speculative, I have suggested that the emergence of human imagination must be continuous with the natural world. More radically, the things of the natural world may move in such a way that this dynamism gives rise to the imagination. Peirce's speculative claim is important in our investigation of the imagination, for it encourages us to consider the ontological implications of the process and its unique dynamics. It also encourages us to investigate the natural basis of the imagination by way of the empirical sciences. Inspired by Peirce, I have employed research from contemporary neuroscience in order to identify the neural basis of imaginative behavior, behavior that establishes conceptual unity, meaning, and novelty. First, I examined the research on cross-modal mapping that underpins metaphoric and creative thinking. This research indicates that certain forms of abstract conceptualization are grounded in sensory-motor experience. The studies of the mirror neuron system indicate that neural coordination of the premotor cortex grants the possibility of action recognition. More simply, these studies point to biological mechanisms that allow human inquiry to coordinate with and *mirror* our social and environmental surroundings. Similarly, the work of Donald Tucker confirms the character of the imagination as described by Peirce and Kant. He explains that human conceptualization is developed by a complex process of neural differentiation and integration and involves visceral activation and more differentiated forms of neural coordination.

Shifting levels of analysis, I addressed the fact that cross-modal coordination depends on the synchronization and the parallel processes of neural populations and began to explore the population dynamics studied by Gerald Edelman. Edelman's theory of neural Darwinism and, more specifically, his concepts of reentry and degeneracy expand the scope of imaginative novelty and stability. At the level of neural population dynamics, Edelman discovers natural processes and characteristics that underpin the creative movements that have long been associated with human artistic imagination. In so doing, he helps us reframe Peirce's assertion that all matter is really partially deadened mind and investigate the emergence of the imagination from the physical world.

The Value of Impossible Projects (The Sphinx Remembered)

A reader might wonder how such a little book can claim so much—claim to explore the intricacies of aesthetic experience and, *at once,* to examine the particular ways in which the emergence and creativity of nature is continuous with this experience. I have often wondered this myself. I placate these worries with two beliefs that I have come to in the writing of this book.

First, I believe that doubt is the only appropriate response to certain questions. What is the nature of the imagination? The answer to this question cuts in two directions. On the one hand, we must attend to the subjective quality of the experience of the imagination; on the other, we must address the objective characteristics of natural processes from which this experience emerges. On the one hand, we approach the question as a

196 BE IMAGINATIVE! SUGGESTION AND IMPERATIVE

phenomenologist; on the other, a cognitive neuroscientist. There is a necessary difference between these two methods of inquiry. It is necessarily frustrating for anyone who attempts to overcome this divide.

I was slower in coming to the second belief, namely that philosophy need not be in the sole business of laying out definitive claims and exhaustive systems. The claim set out in this book should be regarded as a plan for action, vague enough to permit latitude in what action to take, definite enough to motivate future research. Some might say that this approach stands to leave things unfinished or open to criticism on the basis of the fledgling nature of the project. To these readers, I would direct their attention to Peirce and suggest that some projects are better left expressed but unfinished.

The reason this book focuses on Peirce's philosophy instead of that of Dewey or James is that Peirce's writing repeatedly reflects the inherent difficulty of explaining the nature of creative experience. Long before there was a "hard problem of consciousness," Peirce was frustrated by it. The false starts and second guesses that define his corpus are, at least in part, a product of his attempt to *describe subjective experience objectively*. (That is, by the way, what I take Kant's philosophical project to be about as well.) Peirce attempts to demonstrate, exemplify, explicate, graph, quantify, infer, and scrutinize something that defies objective explanation: creative experience. Readers of Pierce are all too familiar with the way in which he begins in a detailed, almost poetic, rendering of phenomenal experience, only to veer wildly into the most arcane analysis of logic and the empirical sciences. More often than not, he cuts himself short after a few pages, as if he has decided that saying more would do little to address the futility of the project he has undertaken. To an outside observer, this appears to be some sort of confused self-sabotage, but to anyone who has thought about the "hard problem" for any amount of time (and now, you too have thought about it for at least a few hours), this veering form of argument that ends abruptly in a moment of frustration is understandable if not instructive. At the very least, Peirce shows us the difficulty—nay, impossibility—of explaining on no uncertain terms the nature of the creative experience. As Colin McGinn explained many years later, this subjective experience can never be explained in any definitive way by empirical observation. The cognitive tool we use in empirical investigations, namely perception, is ill-suited to elucidate the nature of phenomenal experience. In McGinn's words: "Conscious states are simply not the potential objects of perception: they depend on upon the brain but cannot be observed by directing our senses onto the brain."[2] The radical and sudden jumps between Peirce's phenomenological descriptions and his empirical investigations seem to indicate that he was well aware of this fact. Indeed, it is very likely that these dramatic turns in Peirce's argumentation were intended to make *us* aware of this fact. What is remarkable about Peirce's failed attempts to bridge phenomenology and empirical investigation is their tenacious frequency over a period of nearly fifty years. Failure, for Peirce, does not seem to be a good reason to stop the investigation. Instead, failure serves as a sign that more work needs to be done, that one has succeeded in asking a very hard question that appears impossible, at least for the day. Even good answers seem woefully inadequate. If Peirce is right in "The Fixation of Belief," the experience of temporary impossibility, what most people call doubt, initiates inquiry. It is only when we achieve some stable—better yet, indefensible—belief that the inquiring mind can go on holiday. And summer holidays can be so very boring. Or at least they were for Peirce.

Peirce's repeated attempts to provide an account of creative experience, through his development of abduction and synechism, were meant to counteract a growing trend in twentieth-century philosophy to satisfy itself with modest questions that could be *fully* and *definitively* answered with analytic rigor. Peirce believed that this trend was somewhat misguided and risked selling philosophy's possibilities short. Indeed, the problem that he had with both Royce and Dewey at the turn of the century reflects this position. According to Peirce, Royce failed to grasp the importance of spontaneity, contingency, and creativity that defined human cognition; his love of systematizing often led Royce away from the concrete quality of imaginative experience. Dewey attended to the qualitative dimension of experience but failed take seriously Peirce's emphasis on continuity and generality. This is to say that Dewey did not endorse an important aspect of Peirce's metaphysics, or if he did endorse it, he certainly did not highlight it in his writing. The difference between Peirce and Dewey is no greater than that between Peirce and Royce, but for the purposes of the current project a final comment concerning the use of Dewey is warranted.[3]

Over the course of this book, I have repeatedly employed Dewey's treatment of the imagination. Doing so may lead a reader to think that I endorse his position over Peirce's. This is not the case. Dewey provides occasionally beautiful prose to describe one side of Peirce's philosophy of the imagination, but only one side. Dewey describes the coordination between the self and the Universe through the processes of the imagination in his later works, especially in *Art as Experience* and *A Common Faith*, at a time when he was rereading many of Peirce's early writings. Let us return to a comment along these lines that was addressed earlier:

> The idea of a thoroughgoing and deep-seated harmonizing of the self with the Universe (as a name for the totality of conditions with which the self is connected) operates only through the imagination . . . which is one reason why the composing of the self is not voluntary in the sense of an act of special volition or resolution.[4]

Dewey closes *A Common Faith* (1934) with this remark about the way that the harmonization of the self with the Universe occurs only through the creative experience of the imagination. He recognizes that the imagination overcomes the sharp divide between subject and object, that this process is not the same as will or deductive logic, and that it is an embodied process that brings us into real and meaningful touch with the wider world. At first glance, this appears all very Peircean. Dewey, however, downplays what for Peirce was the metaphysical upshot of this union between the self and Universe. For Peirce, it was not merely the case that human imagination draws the self into intimate contact with the world but rather that the world "has its being in a logic of events" that can properly be described as creative or imaginative. In other words, the dynamic of the human imagination is isomorphic with (has the same form as) the dynamic processes of the natural world. These natural processes provide the limiting and enabling conditions for the highest forms of human creativity. Dewey returned to the "hard problem" of Peirce's philosophy relatively late in his career but remained unwilling to take on board the metaphysical commitments Peirce's philosophy entailed. On March 12, 1935, more than two decades after Peirce's death, Dewey reflects on his indebtedness to Peirce, writing to Read Bain: "I could not understand Peirce until I had reached under other influences a position somewhat akin to his. So it is only in later years that his particular work has counted very much with me. Now, I probably depend on it more than upon the work of any other one man."[5] This being said, it is clear from subsequent letters that Dewey did not ascribe to the metaphysics that Peirce was committed to. On February 10, 1939, Dewey received a letter from James Feibleman that pushed him to clarify the issue concerning the relationship between Dewey's pragmatism and Peirce's pragmaticism.

February 10, 1939

Dear Professor Dewey:

May I be pardoned for intruding thus upon your privacy? I met you some years ago at Jacob Epstein's apartment on the occasion of his American visit, but you will not remember me. There is a sense in which no philosopher can escape from the routine of public attentions because he is in fact a public figure, and it is rather to this sense that I appeal in writing to you now.

I am engaged on a study of the philosophy of Peirce. There is to be a last chapter devoted to a discussion of the influence of Peirce, and it is in the latter connection that I have associated your name. I note your acknowledgments to Peirce in different places in your recent Logic, particularly in the general statement in the Preface on page iv: "with the outstanding exception of Peirce, I have learned most from writers with whose positions I have in the end been compelled to disagree." The debt to Peirce which you acknowledge is clear in the text; it is based on methodological and procedural points which I find that you and Peirce do share.

But Peirce's methodology is intimately connected with an ontological position which I am afraid, judging from your work, you would completely repudiate. Peirce in no uncertain terms declared himself "a realist of the Scotian variety," and believed that science is essentially realistic. This not in one passage but in many which leave no doubt as to his metaphysical position.

Would it be asking too much if I requested that you write me, in terms which I could quote over your name, your position with regard to Peirce's realism? I feel that although my own work may not be important, the American Philosophical tradition is, and that the relations between your philosophy and the philosophy of Peirce should be made clear.[6]

Dewey responded to Feibleman with a very brief reply several weeks later. It seems likely that Dewey's response would have been longer had he known that Feibleman would go on to write one of the first major introductions to Peirce's philosophy and that the issue of ontological realism would divide the field of pragmatism for the next half century. But Dewey responds very quickly:

> Dear Mr. Feibleman—
> Yours of the 10th has reached me here.
> I'm sorry I cant make a statement for publication on the point you mention because I think Peirce's logical position is separable from his ontological one, & in addition the latter is stated by him in many different ways—at times he says he is a "Realist" in the medieval sense; in other places he comes close to calling himself a Kantian; in places his metaphysics is definitely panpsychic—And this doesn't exhaust the list. Consequently I am content to take him for what he distinctively was, a pragmatist (or "pragmaticist"), which position is necessarily to my mind realistic—in opposition to idealism—[7]

It is obvious that Dewey missed the thrust of the Feibleman's question. Feibleman was asking Dewey to explain the difference between his version of pragmatism and Peirce's. Peirce repeatedly stated that he was not a pragmatist of the sort that Dewey had in mind. More specifically, Feibleman was asking Dewey to explain how the logical position of pragmatism could be "separable" from ontological realism. This is a claim that Peirce would never make and is what separated Peirce's philosophy from Dewey's pragmatism. What is most striking about Dewey's letter to Feibleman is his willingness to overlook the *many* attempts that Peirce made to define his metaphysical position—at times "realist," at others Kantian, at still others panpsychic. Indeed, Dewey seems to dismiss Peirce's metaphysics on the basis that Peirce tried so often, in so many varied ways, to articulate his stance and to bring it into the proper relationship with his well-articulated logic and epistemology. In light of Peirce's struggle, Dewey is "content to take [Peirce] for what he distinctively was, a pragmatist." Such contentedness, I would maintain, is not the stuff of good inquiry. It also tends to lead to a type of boredom that is antithetical to imaginative genius. Peirce may have been many things—arrogant, insecure, disingenuous—but he was *never* content. After all, Peirce, unlike Dewey, was deeply and sincerely committed to a philosophical project that amounted to squaring the circle. As Maurice Merleau-Ponty says of "Cezanne's Doubt," "expressing what *exists* is an endless task."[8] It is for this reason that my analysis has focused almost exclusively on Peirce's philosophy; Peirce was happy to experience the difficulty of thinking through the imagination. For individuals immersed in their daily affairs, the imagination and genius emerge in the midst of the humdrum, often in response to some difficulty or friction, and transform the world in creative and adaptive ways. For philosophers (and yes, believe it or not, they too are individuals immersed in their daily affairs), the concepts of genius and imagination crop up exactly at the point where philosophical speculation is most difficult, where analytic rigor tends to fail us, where gesturing seems like the most appropriate response to a question. This is not a point to abandon analytic rigor but to redouble our efforts. The Sphinx remains at the gates, but we can give the riddle our best shot. One can easily see this in the way that the imagination and genius serve as placeholders for Kant when he desperately wants to bridge understanding and sensation, or originality and common sense, or subject and object. Peirce's intent was to *explain* in exact detail the nature of the placeholder. And this is an endless and exacting task.

The exactitude of Peirce's thought might occasionally strike us as tedious or unnecessary. Peirce, however, seemed to understand analytic and scientific rigor as the *only* appropriate complement to the metaphysical and cosmological position that he proposed. The closing comment of Peirce's 1898 Cambridge lectures that the world "lives and moves and has its being in the logic of events" is explicitly religious, drawing from Acts 17:28 ("For in him we live and move and have our being"). This religiously inspired remark, however, was not, for Peirce, the call to Christian faith but rather an invitation and challenge to scientists, mathematicians, and logicians to investigate the particular contours of the reasoning and logic of *things*. Peirce accepted this invitation wholeheartedly, filling the years that followed the Cambridge lectures with his studies of continuity. During this time he came to see, repeatedly, that the concept of continuity suggested a metaphysics that could not be exhaustively analyzed but that this inexhaustibility was an opening to cultivate the most rigorous modes of analysis. His methods of investigations became more rigorous even as his metaphysics became more speculative. This is the correlation that defined Peirce's work on the whole, and it is one that Dewey never quite understood. In the midst of his studies on true continua in 1904, a frustrated Peirce lashed out at his philosophical community, defined by Dewey's budding instrumentalism, for its willingness to avoid difficult metaphysical questions and to shun the technical modes of inquiry that might shed added light on them. In a notebook on the nature of ordinals, Peirce writes: "As for the whole existing race of philosophers, say John Dewey, to mention a relatively superior man whom you see, why they are the sort of trash who are puzzled by Achilles and the Tortoise! Think of trying to drive any exact thought through such skulls! Royce is the only philosopher I know of real power of thought now living."[9] This was just months after the Philosophical Conferences were held at Royce's Irving Street home. Despite being Royce's next-door neighbor, James attended only one meeting of the conferences, which were held fortnightly for the better part of a year. At this point, James gave a paper on Dewey's instrumentalism, triumphantly announcing that a genuine school of pragmatism had been founded in Chicago. According to James (in a letter to F. C. S. Schiller, dated November 15, 1903), Royce's system was "cut under" by the report that he had given and was forced to "demean himself in an admirably amicable and candid way" on that evening.[10] Royce may have been amicable on the surface, but I suspect that he was profoundly dismayed that a philosophy such as Dewey's, which seemed to take the path of least resistance—both metaphysically and analytically—could gain such wide appeal. Peirce was privy to the intellectual cockfights of Harvard and, at this point, seemed to stand squarely on Royce's side of the debate. Dewey's pragmatism might have achieved popularity, but according to Peirce and Royce it had sacrificed intellectual integrity and philosophical strenuousness in the process. This is a lesson that we would do well to keep in mind as pragmatism comes into vogue once again.

What does this discussion have to do with the current analysis of the imagination? Quite a bit, as it turns out. It follows that the study of

the imagination, if it is conducted in the spirit of philosophical strenuousness, will not remain at the level of the aesthetics, cultural critique,
or phenomenology—where Dewey tended to leave the issue—but venture into the most complex fields of general biology, mathematics, and
logic; this is where Peirce believes that the nature of the imagination
leads. And yes, this pursuit has metaphysical and religious implications,
but not the sort that I can address fully here. That being said, a few comments can be made to explore these implications a bit further.

The Ground Re-Turned: Tending to the Imagination

In his work on Susanne Langer's conception of art and embodiment,
Donald Dryden does a very nice job of describing the relation between
imaginative consciousness and biological comportment, a relation at the
center of this book. "Consciousness," Dryden writes, "is not a peculiar
product of biological processes—a curious emanation whose relation
to physical events in the central nervous system is merely *contingent*."[11]
Langer reiterates this point on the emergence of consciousness:

> It is a misconception to think of sentience as something *caused* by
> vital activities. It is not an effect, but an aspect of them. . . . Sentience
> arises in vital functioning rather than from it; life as such is sen-
> tient. . . . Naturally, then their basic forms are vital forms; their com-
> ing and going is in the pattern of growth and decline, not of mechanical
> occurrences; their mutual involvements reflect the mold of biological
> existence.[12]

Instead of understanding human consciousness as a *result* or *outcome* of
our embodiment, Langer seems to suggest that it ought to be regarded as
a phase or aspect of biological occurrences and relations: it emerges in
the thick of things. To say that human thought and organic life are continuous is to say that the patterns of imaginative consciousness emerge
in the biological rhythms of the natural world. Such a suggestion ought
to re-turn and revise the field of study that concerns itself with the shape
and function of human cognition. More immediately, if not more importantly, this suggestion ought to return us to our experience as embodied and creative organisms and serve as an exhortation to attend to
our being imaginative.

Peirce often admitted that he could never shake his Transcendental inheritance, the fact that he was brought up alongside of Emerson and Thoreau. Contemporary treatments of Peirce's philosophy, however, attempt to separate it off from these earlier thinkers, in effect feigning to do something that Peirce said explicitly could not or should not be done. This says more about our unwillingness to acknowledge the philosophical in the poetic and the religious than about Peirce's. There are, undoubtedly, significant differences between Peirce and the Transcendentalists, but to forget the points of contact (and there were many) undercuts Pierce's hope that his philosophy would have metaphysical, aesthetic, and ethical implications. It thus seems appropriate to return to a thinker whose work resonates closely with Peirce's and allows us to shift our angle of vision on the current study of the imagination.

Henry David Thoreau's *Walden* is sprinkled with poetic renderings of the emergence and complexity of the natural world, a world I have attempted to explore in the preceding chapters and one I suggest is continuous with human creativity. Thoreau writes *Walden* to awaken us to the potentialities that are too often overlooked in the habits of our civilized and scholarly pursuits. He does so by grounding us in the natural world and by reminding us of our organic and partially wild histories. I have undertaken the writing of this book with a similar goal in mind and have attempted to adopt a similar method. In underscoring the place of the imagination, I have aimed to highlight the creative, aesthetic, and embodied aspect of cognition that has often been overlooked in modernity's thin treatment of the field of human cognition. This oversight, I believe, is a reflection of modernity's more basic tendency to neglect the potentialities and creative movements that might animate the lives of its individuals and institutions. It is with this suspicion in mind that I have concentrated my energies on the examination of the processes of the imagination to re-turn a modest plot of the scholarly field.

Thoreau begins to work through his "Bean-Field" with a unique attention to the creative force of nature and the continuity between this force and the invigorating thinking of human imagination. He writes:

> Meanwhile my beans, the length of whose rows, added together, was seven miles already planted, were impatient to be hoed...

indeed they were not easy to be put off. What was the meaning of this so steady and self-respecting, this small Herculean labor, I knew not. I came to love my rows, my beans, though so many more than I wanted. *They attached me to the earth, and so I got strength like Antaeus. . . . What shall I learn from beans or beans from me?*[13]

Thoreau, and to a lesser extent Peirce, begins to present the way in which organic life beckons to us in creative moments of investigation and the ways in which we are *naturally attuned* to this beckoning. He explains, "It was a singular experience that long acquaintance which I cultivated with beans. . . . I was determined to know beans."[14] "To know beans," Thoreau suggests, is not some wholly abstract or esoteric pursuit. Indeed, it requires an "intimate and curious acquaintance" that can flourish only when tended with a care that takes into account the creative purposes and interests of the world beyond ourselves. More accurately, such acquaintances grow only when we pay attention to the continuity and dependence that define these creative situations; they grow only when I am willing and able to dirty my hands, Antaeus-like, in the thick of things. This creative involvement, emerging from natural interactions and relations, is precisely the purpose of the imagination.

In "The Bean-Field," Thoreau suggests that our "civilized" conception of the aesthetic imagination has become characterized by stultified ways of thinking and acting. The artistic *Spiel* of the imagination has been confined—caged in the staid rooms of museums and our other modern-day salons. To combat the deadening effects of this trend, Thoreau encourages us to return to the natural origins of our creative wellsprings, stating: "Those summer days which some of my contemporaries devoted to the fine arts in Boston or Rome and others contemplation in India, and others to trade in London or New York, I thus, with other farmers of New England, devote to husbandry."[15] In the midst of a type of husbandry—in a practice that engenders and extends the continuity of meaning and life—is the *dwelling* of the imagination. Husbandry derives its meaning from the verb *búa*, "to dwell." A careful dwelling, cultivating, and flourishing is precisely the provenance of the imagination hitherto described. It is this dwelling in which nature itself continually and spontaneously makes its home.

Thoreau tells us that "ancient poetry and mythology suggest that husbandry was once a sacred *art*; but it is pursued with irreverent haste and heedless by us, our object being to have large farms and large crops merely."[16] He entreats us to return the husbandry of life—our life—to its rightful status as a sacred art. In so doing, it seems appropriate and indeed necessary to move with reverent patience in thinking through the imagination, in participating in these processes that mediate between ourselves and the natural world. Thoreau anticipates this sentiment when he refers to his bean field, writing that "mine was, as it were, the connecting link between wild and cultivated fields; as some states are civilized, and others half-civilized, and others savage or barbarous, so my field was, though not in a bad sense, the half-cultivated field."[17] Wildness and spontaneity still emerge in this field, although they emerge within certain natural constraints. Thoreau wishes to preserve these half-cultivated fields and dispositions, for it is in being half-cultivated that we might be fully human. My intuitions concerning the nature of "being imaginative" are spoken in a similar voice (albeit far less eloquent): by concentrating on the creative, continuous, and spontaneous character of cognition, we investigate the half-cultivated aspect of human thinking. It is an aspect of thinking that serves as the point of contact between the ostensibly human and the ambiguously natural. It is an aspect of thinking that deserves more attention if we are not to forget the latent potentialities that silently ground our lives.

What makes husbandry, this creative art of dwelling, a *sacred* art? What makes Kant suggest that reflective judgment is rooted in an "art of the human soul?" What allows Peirce to suggest that creative abduction might be a powerful but too long neglected argument for the sacred character of life? Surely, it is not merely the fruits of the labor that earn these processes of the imagination their unique designation. Thoreau explains that it is the mindful and creative participation that husbandry invites (rather than its acquisition of fruits) that gives this art its sacred character. This active participation, demonstrated in Kantian genius, in Peircean musement, in Thoreauvian husbandry, in Edelman's accounts of reentry, turns on the ability to coordinate in novel ways with wider environmental situations. In the thoroughgoing and deep-seated coordination of the imagination, a child is encouraged to *lose himself* in the creativity that the world affords and, in so doing, occasionally *find himself* as being an integral part of this flourishing garden that surrounds him. This imaginative child is encouraged to *sacrifice* something of his or her habitual life and in so doing *make sacred* this garden that he or she comes to call his own.

The gardener-genius is slow to lay claim to the products of his imagination, for they are, in an important respect, not wholly in his or her possession. After all, according to Thoreau, we learn something from the beans, if we only have ears and eyes for the lesson. Thoreau elaborates on this point, suggesting that the beans have their way with the gardener and have "results" that are not dependent on the gardener's hands:

> This broad field which I have looked at for so long looks not to me as the principal cultivator, but away from me to influences more genial to it, which water and make it green. These beans have results that are not harvested by me. . . . The true husbandman will cease from anxiety, as the squirrels manifest no concern whether the woods will bear chestnuts this year or not, and finish his labor everyday, relinquishing all claim to the produce of his fields, and sacrificing in his mind not only his first but his last fruits also.[18]

Here Thoreau suggests that the imagination that we think through, the imagination that cultivates our respective fields, is never fully in our possession. More accurately, we possess it only to the extent that we recognize the continuity between its creative movement and the generative force of our surroundings. The freedom of the imagination is realized in and arises from the creative workings of nature. Thoreau reflects on this freedom, commenting that "in solitude, I go and come with a strange liberty in Nature, *a part of herself.*"[19] This comment doubles back on itself, making an important point, namely that such liberty "in Nature" never obtains simply "in solitude." Perhaps the young Thoreau, as he tended Emerson's household garden in 1844, came to understand the curious liberty of thinking through the imagination: after a stint in the field, it is nearly impossible to see where the dirt ends and the skin begins. In the thick of things where the imagination emerges, such distinctions are difficult to make.

The Imaginative "Thou"

Thoreau allows us to see that being imaginative involves the recognition that there are interests and creative forces beyond our narrow self-definition, that the world we see and experience has its own creative purposes. It is working on its own behalf. In the words of William James, we come to see in the imagination that "the Universe is not a mere It to us, but a Thou . . . and any relation that is possible from person to person might be possible here."[20] This comment was made in the "Will to Believe" (1896) and anticipates Martin Buber's *I and Thou* (1923) in its depiction of a unique and living relationship between self and world. This I-Thou relationship is living in the sense that it is not one between a subject and object but rather between two subjects that share equally the unity of being. In this relationship a person has the ability to cultivate *and be cultivated by* the beans of a garden. I have introduced James's suggestion concerning the "Thou" for a particular reason, as a suitable closing to our thinking through the imagination.

Where did James get this idea of the "Thou" relationship that transcended the subject-object dualism? He gives us a hint in an passing remark in 1862: "The *thou* idea, as Pierce calls it, dominates an entire realm of mental phenomena, embracing poetry, all direct intuition of nature, scientific *instincts*, relations of man to man, morality &c. . . . *All analysis* must be into a triad; *me & it* require the complement of *thou*."[21] The "thou idea," like so many others that have made their way into the writing of other American thinkers, was, at least at first, Peirce's. In the spring of 1861, Peirce had begun a book entitled *I, IT, and THOU*. He writes that, "I here, for the first time, begin a development of these conceptions. . . . THOU is an IT in which there is another I. I looks in, It looks out, Thou looks *through*, out and in again."[22]

I remember quite vividly the summer afternoons of childhood boredom when my brother and I kept busy by bickering or tearing up, instead of planting, my mother's garden. That is to say, we regarded the world and each other in a sort of distant third-person perspective— alien, cold, inert. *It* did not matter if we trampled the flowers or each other. We could just as well do without *it*. Only in my mother's injunction to "be imaginative" was this alien world made intimate and animate once again. Slowly, I came to attend to and care for little things with petals and roots, to care for them in ways that belied a curious truth, namely that they were not *mere* things. And somehow I always ended up caring most for the littlest plants who had chosen the most unlikely places to grow. They arose unexpectedly and called for different treatment in order to flourish. To recognize this call as the call of another "I" is to enter into a "Thou relationship" that can only be realized by thinking through the imagination. A final remark on Peirce's philosophy seems warranted to make this point clearly.

To suggest that the "Thou idea" was wholly Pierce's is not entirely accurate. Just as James borrowed it from Peirce, Peirce borrowed it from the German idealists mentioned in the opening chapters of this book. Buried in one of his undergraduate essays on Ruskin is Peirce's description of the three impulses outlined by Friedrich Schiller in his *Aesthetic Letters*: *Formtrieb*, *Stofftrieb*, and *Spieltrieb*. Here the young Peirce explains that Schiller intended the play drive (*Spieltrieb*) to mediate and balance between the formal drive (*Formtrieb*) that stems from the rationality of a subject and the sensuous drive (*Stofftrieb*) that proceeds from the physical existence of a human being. Appended to his analysis of these three impulses is an interesting aside that deserves a central place in the current discussion. Peirce writes: "I should say that these were the I impulse and faculty, and the IT impulse and faculty; and also the THOU impulse and faculty which (it seems to me) is what Schiller regards as that of beauty."[23] It is only a short step from the play of beauty described by Schiller to the role of the imagination set forth by Kant in the schematism and reflective judgment. Indeed, the resonances between Kant and Schiller have been addressed in the second and third chapters of this book. Peirce, therefore, suggests that the "THOU impulse" is related in an intimate way to the processes of the imagination that he would later develop as abduction and musement. Only through the imagination are we able to grasp—nay, consider—the IT as another I. "Grasp" is too strong a word; the imagination can never definitively get hold of the existence of a Thou. It does not set the bounds of and exhaustively describe the existence of this relationship. Instead the imagination gestures toward and allows us to consider the reality of the Thou, a reality that remains warmly intimate but forever beyond our reach.

Notes

1. THE CULTIVATION OF THE IMAGINATION

1. The epigraph above is from Charles Peirce, *Collected Papers of Charles Sanders Peirce*, vols. 1–6, ed. Charles Hartshorne and Paul Weiss, vols. 7–8, ed. Arthur W. Burks (Cambridge, Mass.: Harvard University Press, 1960). References to the *Collected Papers* will be given in the usual manner; e.g., CP 6:289. Peirce citations taken directly from the papers at Houghton Library will be given in the usual manuscript manner; e.g., MS 123.

2. John Dewey, *Art as Experience* [1934], in *John Dewey: The Later Works, 1925–1953*, ed. Jo Ann Boydson (Carbondale: Southern Illinois University Press, 1986), 12:277.

3. Ibid.

4. Ibid., 271.

5. Plato, *Republic*, 601.

6. Plato, *Ion*, 534.

7. Mark Johnson, *The Body in the Mind: The Bodily Basis of Meaning, Imagination, and Reason* (Chicago: University of Chicago Press, 1987), 145.

8. John Sallis, *Delimitations: Phenomenology and the End of Metaphysics* (Bloomington: Indiana University Press, 1995), 6.

9. Ibid.

10. The epigraph to this section is from CP 1:46.

11. CP 1:201.

12. Italics mine. Dewey, *Art as Experience*, 278.

13. Douglas Anderson, *Creativity and the Work of C. S. Peirce* (Boston: Martinus Nijhoff, 1987), 6.

14. Christopher Hookway, *Peirce* (London: Routledge, 1985), 155.

15. Charles Peirce, *Reasoning and the Logic of Things*, ed. K. Ketner (Cambridge, Mass.: Harvard University Press, 1992), 161.

16. George Lakoff and Mark Johnson, *Philosophy in the Flesh: The Embodied Mind and Its Challenge to Western Thought* (New York: Harper Collins, 1999).

17. Italics mine. Antonio Damasio, "Some Notes on Brain, Imagination, and Creativity," in *The Origins of Creativity*, ed. K. Pfenninger and V. Shubik (Oxford: Oxford University Press, 2001), 59.

18. Gerald Edelman and Giulio Tononi, *A Universe of Consciousness: How Matter Becomes Imagination* (New York: Basic Books, 2000), 48.

19. Mary Warnock, *Imagination* (Berkeley: University of California Press, 1976), 9.

20. Ibid.

21. Ibid., 10.

22. Douglas R. Anderson and Carl R. Hausman, *Conversations on Peirce* (New York: Fordham University Press, 2012), 34.

23. Ibid., 65.

2. ENLIGHTENING THOUGHT: KANT AND THE IMAGINATION

1. Sections of this chapter appeared in John Kaag, "Continuity and Inheritance: Kant's *Critique of Judgment* and the Work of C. S. Peirce," *Transactions of the Charles S. Peirce Society* 41, no. 3 (2005): 515–540.

2. Plato, *Ion*, 534C.

3. This gap is brought to light by various authors described below. Here, I am employing Eckart Förster's language. Eckart Förster, *Kant's Final Synthesis: Essays on the* Opus Postumum (Cambridge, Mass.: Harvard University Press, 2005), 48–75.

4. Immanuel Kant, *The Critique of Pure Reason* [1781], trans. P. Guyer and A. Wood (Cambridge: Cambridge University Press, 1998); *The Critique of Judgment* [1790], trans. W. Pluhar (Indianapolis, Ind.: Hackett, 1987). The *Critique of Pure Reason* will henceforth be referred to as CR in citations. The *Critique of Judgment* will be referred to as CJ.

5. Immanuel Kant, *Opus Postumum* [1794], trans. E. Förster (Cambridge: Cambridge University Press, 1993); *Anthropology from a Pragmatic Point of View* [1798], trans. M. Gregor (The Hague: Nijhoff, 1974). The *Opus Postumum* will be referred to as OP in citations.

6. Henry Allison and Donald Crawford both make this point in their analyses of Kant's aesthetic. Henry Allison, *Kant's Theory of Taste: A Reading of the* Critique of Judgment (Cambridge: Cambridge University Press, 2001), 123–147; Donald Crawford, *Kant's Aesthetic Theory* (Madison: University of Wisconsin Press, 1974), 78–118; Donal Crawford, "Kant's Theory of Creative Imagination," in *Essays in Kant's Aesthetics*, ed. Ted Cohen and Paul Guyer (1982), 151–178.

7. CR xxii.

8. Kuno Fischer, *A Commentary on Kant's* Critik of Pure Reason (London: Longman and Green, 1866), 2.

9. CR xxiv.

10. CR A79–81.

11. CR B143.

12. Henry Allison, *Kant's Transcendental Idealism: An Interpretation and Defense* (New Haven, Conn.: Yale University Press, 2004), 170.

13. CR A89–90.

14. CR B1.

15. CR A116–128.

16. CP 1.35.

17. MS 1004.

18. CP 1:39.

19. CP 1:39.

20. CR B152–169.

21. CR A135.

22. CR A138.

23. CR A138.

24. CR A138. This chapter falls short for many reasons. One noteworthy reason, not addressed in detail, is the fact that Kant spends most of the chapter dealing with time and "inner sense," without specifying the conditions by which "outer intuition" seized upon particular objects.

25. Here Kant is only referring to the operation of the schema on empirical concepts of the understanding; in the process furnished by transcendental schema that addresses only pure categories, no images are produced.

26. CR A78.

27. CR A147.

28. Italics mine. CR A15.

29. CR A138/B178.

30. Mark Johnson, *The Body in the Mind: The Bodily Basis of Meaning, Imagination, and Reason* (Chicago: University of Chicago Press, 1987), 157.

31. Martin Heidegger, *Phenomenological Interpretation of Kant's* Critique of Pure Reason [1927–1928], trans. P. Emad (Bloomington: Indiana University Press, 1997).

32. This description is given most succinctly in the first chapter of the second book of Heidegger's lectures. Most notable are three interlocking sections: "The Power of the Imagination as the Source of Comprehensive Synthesis," "The Pure Imaginative Time-Related Synthesis as the Source of the Concepts of Understanding," and "The Unity of Imaginative Synthesis and the Unity of the Logical Functions of Judgment." The lengthy titles of these sections are suggestive. Heidegger's remarks illustrate the way in which Kant designates the imagination as a synthesizing power necessary in grounding the categories in experience and in forming a type of unity in the logical function of judgment.

33. Ibid., 278.

34. CP 1:35.

35. Hannah Arendt, *The Life of the Mind* (New York: Harcourt, 1977). For Gadamer's discussion of "play," see Hans-Georg Gadamer, *Truth and Method*, trans. G. Barden (New York: Seabury, 1975), 56–81.

36. John Sallis, *The Gathering of Reason* (Albany, N.Y.: SUNY Press, 2005).

37. Ibid.

38. Paul Guyer, *Kant and the Claims of Taste* (Cambridge, Mass.: Harvard University Press, 1979).

39. Allison, *Kant's Transcendental Idealism*; Henry Allison, *Kant's Theory of Freedom* (Cambridge: Cambridge University Press, 1990).

40. Johnson, *The Body in the Mind*.

41. Makkreel argues that Kant's transcendental philosophy is compatible with modern hermeneutics. He emphasizes the role of the imagination in the process of interpretation, a process that forces a rereading of Kant's discussion of a "feeling of life," common sense, and purposive history. Rudolfe Makkreel, *Imagination and Interpretation in Kant: The Hermeneutical Import of the Cri-tique of Judgment* (Chicago: University of Chicago Press, 1990).

42. Zammito describes the tension reflected in Kant's critical project between Enlightenment rationality and the belief in the creative, imaginative genius that gives rise to German idealism. He underscores the culminating character of the third *Critique* and argues that this work seeks to integrate aesthetics into general epistemology after its exile from Enlightenment philosophy. John Zammito, *The Genesis of Kant's* Critique of Judgment (Chicago: University of Chicago Press, 1992).

43. Henry Allison, *Kant's Transcendental Idealism: An Interpretation and Defense* (New Haven, Conn.: Yale University Press, 2004), 192.

44. Johnson's description of imagination exposes the inadequacy of Kant's epistemology but also the way in which Kantian schematism might be reinter-preted in light of advances in cognitive neuroscience. Johnson, *The Body in the Mind*, 139–172.

45. Förster argues that this "self-positing" that occupied Kant's thought in the closing years of his life is, by its very definition, imaginative and stands as the precondition of cognition. Eckart Förster, "Kant's *Selbstsetzungslehre*," in *Kant's Transcendental Deductions*, ed. E. Förster (Stanford, Calif.: Stanford University Press, 1989), 217–238.

46. Allison provides ample discussion of the Kantian thing-in-itself in his *Kant's Transcendental Idealism*, 50–73. For Kant's description, see CR A251–252, B307, A263/B319.

47. Peirce, as a "Critical Common-Sensist," would later write that "The Kan-tist has only to abjure from the bottom of his heart the proposition that a thing-

in-itself can, however indirectly, be conceived; and then correct the details of Kant's doctrine accordingly, and he will find himself to have become a Critical Common-Sensist."

48. Förster, *Kant's Final Synthesis*.

49. Zammito, *The Genesis of Kant's* Critique of Judgment, 4–5, 151–158, 220–225.

50. CJ iv.

51. In the highly technical introductions to the third *Critique*, Kant presents the *Critique of Judgment* "as mediating the connection between the two parts of philosophy (theoretical and practical) to form a whole." Right away, one is struck by Kant's anticipation of the pragmatic, and particularly Peircean, project. CJ 174.

52. Eva Schaper, "Taste, Sublimity, and Genius. The Aesthetics of Nature and Art," in *The Cambridge Companion to Kant*, ed. Paul Guyer (Cambridge: Cambridge University Press, 1982), 367.

53. Rodolph Gasché, *The Idea of Form: Rethinking Kant's Aesthetic* (Stanford, Calif.: Stanford University Press, 2003), 149.

54. CJ iv.

55. CJ iv.

56. CJ iv.

57. CJ 15.

58. Heidegger, *Phenomenological Interpretation of Kant's* Critique of Pure Reason, 182.

59. This point ought to be underscored in reference to the forthcoming discussion of "self-positing" of the *Opus Postumum*. CJ 22.

60. CJ 9.

61. Donald Crawford, "Kant's Theory of Imagination," in *Essays in Kant's Aesthetics*, ed. Ted Cohen and Paul Guyer (Chicago: University of Chicago Press, 1982), 173.

62. Förster, *Kant's Final Synthesis*, 27.

63. It is interesting to note the family resemblance reflected between Peircean thirdness and aesthetic harmony. Peirce's notion of "thirdness" is a mediating acting that brings distinct entities into relation without destroying their respective particularities. See Makkreel, *Imagination and Interpretation in Kant*, 47.

64. Ibid., 48.

65. Kant's use of terminology in this work helps him describe an experienced immediacy in the play of imagination that is absent in his earlier works. The free play of the imagination is "an occupation agreeable on its own account." It embodies a "purposeless purpose." Kant writes that "the mere formal purposiveness in the play of the subject's cognitive powers . . . is the pleasure itself" that guides the harmonious play of the imagination. All three of these

statements seem to speak to Kant's desire to develop a moment of praxis in relation to aesthetic judgment, in which the ends and constraints of judgment are
implicit in an actual and particular practice of judging. Gadamer appropriates
and extends this notion of play as praxis in his hermeneutics. Gadamer's phenomenological description of play as a hermeneutical moment emphasizes the
self-generating constraints of the game and notes, in Peircean fashion, that inquiry happens in the dynamic movement of spontaneity *and* constraint. Huizinga's analysis of "free play" also seems to make this point. Both Gadamer and
Huizinga extend this pragmatic aesthetic and explicitly elaborate on Kant's
initial claims. Admittedly, his claims do not fully flesh out this movement of
praxis. Both, however, seem to acknowledge fully their indebtedness to the
third *Critique* and recognize the way Kant anticipates many of the moves made
by twentieth-century phenomenology and American pragmatism.

66. Guyer, *Kant and the Claims of Taste*, 118.

67. CJ 63.

68. CJ 170.

69. CJ 170.

70. CJ 170.

71. CJ 173.

72. CJ 66.

73. CJ 94.

74. CJ 94.

75. Drucilla Cornell, "Enlightening and Enlightenment: A Response to John
Brenkman," *Critical Inquiry* 26: 128–140.

76. CJ 94.

77. Hannah Arendt, *Lectures on Kant's Political Philosophy*, ed. Ronald
Beiner (Chicago: University of Chicago Press, 1982), 42.

78. CJ 294–295.

79. Zammito, *The Genesis of Kant's* Critique of Judgment, 139.

80. CJ 161.

81. CJ 203.

82. CJ 203.

83. CJ 307.

84. Jerry Sobel, "Arguing, Accepting, and Preserving Design," in *Essays
in Kant's Aesthetics*, ed. Ted Cohen and Paul Guyer (Chicago: University of
Chicago Press, 1982), 301.

85. CJ 308.

86. CJ 317.

87. CJ 183

88. CJ 199.

89. CJ 199.

90. CJ 280–336.

91. Gasché makes a sustained case for regarding the *Critique of Judgment* as the keystone of Kant's investigation of human inquiry. Gasché, *The Idea of Form*, 42.

92. Ibid.

93. Italics mine. CJ 317.

94. CJ 313–319.

95. Kant writes that the imaginative attunement, one that "quickens" both sense and understanding, is "universally communicable." CJ 239.

96. CJ 307–311.

97. CJ 307–311, 319, 344, 350–351.

98. The form-matter distinction that has characterized much of our discussion of Kant will be drawn into question as the pragmatist investigates the implications of envisioning a more robust role for the imagination.

99. OP 22.8.

100. CJ 173.

101. CJ 146.

102. CJ 146.

103. Förster, *Kant's Final Synthesis*, 11.

104. OP 421.

3. C. S. PEIRCE AND THE GROWTH OF THE IMAGINATION

1. Phillip Weiner, "Peirce's Evolutionary Interpretation of the History of Science," in *Studies in the Philosophy of Charles S. Peirce*, ed. Phillip Wiener and Fredric Young (Cambridge, Mass.: Harvard University Press, 1952), 145. See also Sandra Rosenthal, *Charles Peirce's Pragmatic Pluralism* (Albany, N.Y.: SUNY Press, 1994), 92.

2. While Anderson's reading helped inspire my project, the following arguments diverge in several important aspects from Anderson's account. For now, it seems sufficient to say that Anderson's division of scientific and artistic creativity will not be carried out as an important aspect of the project. The emphasis on the imagination as an epistemological category draws this distinction (between science and art) into question and suggests that imaginative/aesthetic processes are operative in all of human cognition. See Douglas Anderson, *Creativity and the Work of C. S. Peirce* (Boston: Martinus Nijhoff, 1987), 6.

3. Our descriptions, however, diverge in the important sense that I will not argue that Peirce develops an anthropomorphic interpretation of the "raw universe." See Murray Code, "Interpreting the 'Raw Universe,'" *Transactions of the Charles S. Peirce Society* 35, no.4 (1999): 704.

4. Beverly Kent, *Charles S. Peirce—Logic and the Classifications of the Sciences* (Kingston: McGill University, 1987), 150.

5. Ibid.

6. Albert Levi, "Peirce and Painting," *Philosophy and Phenomenological Research* 23, no. 1 (1962): 23–36.

7. Hausman's suggestion that the ordered relations and qualities of a painting point to, and are continuous with, the ordered relations of nature provides an odd echo of Kant's proposal that the freedom of the work of art is commensurate with the freedom of nature. Carl Hausman, "Insight in the Arts," *Journal of Aesthetics and Art Criticism* (Winter 1986): 165.

8. Richard Smyth, *Reading Peirce Reading* (Lanham, Md.: Rowman and Littlefield, 1997), 225.

9. CP 5:xxi.

10. Charles Peirce, "Private Thoughts," in *Writings of Charles Sanders Peirce,* ed. Max Fisch et al. (Bloomington: Indiana University Press, 1982), 1:4.

11. Charles Peirce, "The Sense of Beauty," in *Writings of Charles Sanders Peirce,* ed. Max Fisch et al. (Bloomington: Indiana University Press, 1982), 1:11.

12. MS 1118. n.d.

13. MS 1118.

14. MS 1118.

15. Ibid.

16. MS 1118.

17. MS 1402.

18. As cited in Milton Nahm, "The Imagination as the Faculty for Creating Another Nature," in *The Proceedings of the Third International Immanuel Kant Conference* (Dordrecht: Reidel, 1972), 449.

19. Peirce's translations can be found in full in MS 1005–1007. Smyth maintains that Peirce read the *Critique of Judgment* first, but this claim seems to run against the documentary evidence that shows him translating the first *Critique* at the age of nineteen with his aunt.

20. CP 1:112. Peirce notes that Kant inspired his development of the word pragmaticism. Referring to himself as "the author," Peirce writes: "It was from Kant that the author learned to philosophize. He was soaked with Kant, and was thinking of his word *pragmatische* in the *Metaphysik der Sitten* as meaning purposeful; and what he wanted the word to suggest was that the doctrine [of pragmaticism] made purpose an essential element of rational meaning." MS 291.

21. MS 349.

22. CP 1:240–256. Abbot, Peirce's partner in his early studies of Kant, restates this dissatisfaction at length in his 1906 *Syllogistic Philosophy.*

23. Francis Abbot, *The Syllogistic Philosophy* (Boston: Little Brown and Co., 1906), 36.

24. CP 1:35.

25. CP 1:46.

26. MS 848.

27. Carl Hausman, *Charles S. Peirce's Evolutionary Philosophy* (Cambridge: Cambridge University Press, 1993), 57.

28. CP 1:546.

29. CP 1:548.

30. CP 2:225.

31. "Synechism" is discussed in reference to imaginative continuity in chapter 4.

32. CP 2:238. See also James Feibleman, "Peirce's Use of Kant," *Journal of Philosophy* 42, no. 14 (1945): 365–377.

33. Charles Peirce, "Thinking as Cerebration," in *Writings of Charles Sanders Peirce*, ed. J. Kloesel (Bloomington: Indiana University Press, 1986), 4:45. See also MS 288, in which Peirce takes on Kant's understanding of the thing in itself: "The *Ding an sich* can neither be indicated or found. Consequently nothing true or false can be said of it. Therefore, it is nonsense and all that Kant says in reference of it must be set down as meaningless. Excuse this oversight, and we all can see that Kant regards space, time and the categories just as anyone does and never doubted their objectivity."

34. Ibid., 4:46.

35. CP 5:284.

36. CP 5:284.

37. Rodolphe Gasché, *The Idea of Form: Rethinking Kant's Aesthetic* (Stanford, Calif.: Stanford University Press, 2003).

38. CP 5:289.

39. CP 5:311.

40. Carl Hausman, *Charles S. Peirce's Evolutionary Philosophy* (Cambridge: Cambridge University Press, 1993), 14.

41. MS 290.

42. CP 7:361.

43. For a description of probabilistic inference in these articles, see Joseph Esposito, *Evolutionary Metaphysics* (Athens: Ohio University Press, 1980), 133–151.

44. CP 2:647.

45. Aristotle, *De Anima* (Harthworth: Penguin, 1986).

46. CP 2:647.

47. Douglas Anderson, "The Evolution of Peirce's Concept of Abduction," *Transactions of the Charles S. Peirce Society* (Spring 1986): 149.

48. See W. M. Brown, "The Economy of Peirce's Abduction," in *Transactions of the Charles S. Peirce Society* 19, no. 4. (1983): 398.

49. Peirce makes this distinction earlier, in the Lowell lectures of 1866, but does not emphasize its significance. CP 6:609.

50. CP 2:642.

51. Douglas Anderson, "Peirce on Metaphor," *Transactions of the Charles S. Peirce Society* 20, no. 4 (1984): 466.

52. CP 2:643.

53. Thomas Alexander, "Pragmatic Imagination," *Transactions of the Charles S. Peirce Society* 26, no. 3 (1990): 328.

54. CP 7:219.

55. Parentheses mine. A. J. Ayer, *The Origins of Pragmatism: Studies in the Philosophies of Charles Sanders Peirce and William James* (San Francisco: Freeman, 1968), 78.

56. Ibid., 43.

4. ABDUCTION: INFERENCE AND INSTINCT

1. MS 1118.

2. Jaakko Hintikka, "What Is Abduction? The Fundamental Problem of Contemporary Epistemology," *Transactions of the Charles S. Peirce Society* 34, no. 3 (1998): 503.

3. Harold Frankfurt, "Peirce's Notion of Abduction," *Journal of Philosophy* 55 (1958). Reilly suggests that Peirce continues to confuse abduction with a form of induction. See F. E. Reilly, *Charles Peirce's Theory of Scientific Method* (Bronx, N.Y.: Fordham University Press, 1970).

4. John R. Josephson and Susan G. Josephson, eds., *Abductive Inference: Computation, Philosophy, Technology* (Cambridge: Cambridge University Press. 1994).

5. CP 2:623.

6. Douglas Anderson, "The Evolution of Peirce's Concept of Abduction," *Transactions of the Charles S. Peirce Society* 22, no. 1 (1988): 148.

7. Ibid., 150.

8. CP 5:189.

9. Peirce employs this phrasing in the "The Neglected Argument for the Reality of God," in which he claims that abductive methods have been overlooked in the historical attempts to prove God's reality.

10. CP 5:602.

11. Kapitan notes that this type of adoption should be held apart from the character of belief or acceptance. See Tomas Kapitan, "Peirce and Structure of Abductive Inference," in *Studies in the Logic of Charles Sanders Peirce*, ed. Nathan Houser et al. (Bloomington: Indiana University Press, 1997), 487.

12. Douglas Anderson, "The Evolution of Peirce's Concept of Abduction," *Transactions of the Charles S. Peirce Society* 22, no. 1 (1988): 154.

13. CP 5:181.

14. Frankfurt, "Peirce's Notion of Abduction," 594.

15. CP 5:181.

16. CP 5:181.

17. MS 856.

18. CP 5:172.

19. Thomas Kapitan, "Peirce and the Structure of Abductive Inference," in *Studies in the Logic of Charles Sanders Peirce*, ed. Nathan Houser et al. (Bloomington: Indiana University Press, 1997), 489.

20. CP 4:56.

21. W. V. O. Quine, "Epistemology Naturalized," in *Ontological Relativity and Other Essays* (New York: Columbia University Press, 1969).

22. CP 5:604.

23. CP 6:491.

24. Kapitan, "Peirce and Structure of Abductive Inference," 482.

25. CP 5:60.

26. CP 7:223.

27. Hintikka, "What Is Abduction?" 512.

28. Ibid., 514.

29. Fredrick Will, *Beyond Deduction* (New York: Routledge, 1988), 34.

30. Hintikka in "What Is Abduction?" concedes that there may be a hidden theory of strategy in Peirce's concept of habit formation.

31. Josephson and Josephson, eds., *Abductive Inference*, 200.

32. Ibid., 205.

33. Arthur Burks, "Peirce's Theory of Abduction," *Philosophy of Science* 13 (1946): 303.

34. Charles Peirce, *Reasoning and the Logic of Things*, ed. K. Ketner (Cambridge, Mass.: Harvard University Press, 1992), 193.

35. Charles Peirce, "Logical Machines," in *Writings of Charles Sanders Peirce*, ed. N. Houser (Bloomington: Indiana University Press, 2000), 6:70.

36. Gilbert Harman, "Inference to Best Explanation," *Philosophical Review* 74 (1965): 88–95.

37. Lorenzo Magnani, *Abduction, Reason, and Science: Processes of Discovery and Explanation* (New York: Kluwer, 2001), 19.

5. IMAGINING NATURE

1. The epigraph to this chapter is from Friedrich Schiller, *On the Aesthetic Education of Man*, ed. and trans. E. Wilkinson and L. Willoughby (Oxford: Clarendon, 1967), 219.

2. CP 2:197.

3. MS S80. This "Anna Lowell" most likely refers to Anna Cabot Lowell (1818–1894), a close friend of the family of Benjamin Peirce. She published *Seed Grains of Thought* (1856), a collection of poetry including Schiller and cites

B. Peirce's comments on God and divinity. B. Peirce also reviews and recommends her 1846 *An Introduction to Geometry and the Science of Form*.

4. MS S80.

5. MS S80.

6. "*Nichts ist frei in der Natur, aber auch nichts ist willkürlich in derselben.*" Friedrich Schiller, *Kallias*, in *Classic and Romantic German Aesthetics*, ed. J. Bernstein (Cambridge: Cambridge University Press, 2003), 167.

7. Jeffrey Barnouw, "'Aesthetic' for Schiller and Peirce: The Neglected Origin of Pragmatism," *Journal of the History of Ideas* 49, no. 4 (1988): 607–632.

8. As quoted in ibid., 627.

9. John Kaag, "Continuity and Inheritance: Kant's *Critique of Judgment* and the Work of C. S. Peirce." *Transactions of the Charles S. Peirce Society* 41, no. 3 (2005): 515–540.

10. Karl-Otto Apel, "From Kant to Peirce: The Semiotical Transformation of Transcendental Logic," in *Proceedings of the Third International Kant Congress*, ed. L. Beck (Dorstrecht: Reidle, 1972), 106. Similar arguments, based wholly on analyses of the first *Critique*, are made in Kevin Decker, "Ground, Relation, and Representation: Kantianism and the Early Peirce," *Transactions of the Charles S. Peirce Society* 38, no. 2 (2001): 199; and Chandra Rosenthal, "A Pragmatic Appropriation of Kant: Lewis and Peirce," *Transactions of the Charles S. Peirce Society* 38, no.1 (2002): 253–266.

11. Justus Buchler, *Charles Peirce's Empiricism* (London: Keagan, 1939), 257.

12. CP 6:459.

13. In the "Neglected Argument," Peirce deepens his newfound interest in the aesthetic, as demonstrated in his 1903 Harvard lecture "The Reality of Thirdness," in which Peirce encourages his audience to obtain "the aesthetic state of mind" which he says is a "pure state of feeling." At another point in the same lecture, Peirce describes this aesthetic feeling as "a sort of intellectual sympathy, a sense that here is a feeling that one can comprehend, a reasonable feeling." See CP 5:111.

14. CP 2:270.

15. CP 5:494.

16. CP 5:171.

17. Charles Peirce, *Reasoning and the Logic of Things*, ed. K. Ketner (Cambridge, Mass.: Harvard University Press, 1992), 111.

18. CP 6:10.

19. CP 1:383.

20. CP 5:604.

21. CP 4:27.

22. CP 1:473.

23. Bruce Kuklick, *A History of Philosophy in America, 1720–2000* (Oxford: Oxford University Press, 2001), 158.

24. CP 6:388.

25. David Savan, "Peirce and Idealism," in *Peirce and Contemporary Thought*, ed. Ken Ketner (Bronx, N.Y.: Fordham University Press, 1995), 317.

26. MS 735.

27. David Schum, "Evidence Marshalling for Imaginative Fact Investigation," *Artificial Intelligence and Law* 9 (2001): 181. Emphasis added.

28. Douglas Walton, *Abductive Reasoning* (Tuscaloosa: University of Alabama Press, 2004), 207. Emphasis added.

29. W. M. Brown, "The Economy of Peirce's Abduction," in *Transactions of the Charles S. Peirce Society* 19, no. 4 (1983): 400.

30. To a large extent, I have employed Ketner's analysis in order to unpack the Alpha system. See Kenneth Ketner, *Elements of Logic: An Introduction to Peirce's Existential Graphs* (Lubbock, Tex.: Arisbe Associates, 1996), 34–46. For Beta and Gamma graphs, see Fernando Zalamea's outstanding treatment of them in Fernando Zalamea, "Peirce's Logic of Continuity: Existential Graphs and Non-Cantorian Continuum," *Review of Modern Logic* 9, no. 1 (2003): 115–162.

31. MS 280.

32. MS 905.

33. Cited in Kelly Parker, *The Continuity of Peirce's Thought* (Nashville, Ky.: Vanderbilt University Press, 1997), 66. Note that Peirce's investigation of EG also can be interpreted as pointing forward into a study of his semiotic. Peirce believes that a diagram is to be understood as an icon that gives access to the relations and forms that constitute human reasoning.

34. MS 293.

35. Ketner, *Elements of Logic*, 31.

36. MS 280.

37. Ketner, *Elements of Logic*, 32.

38. Peirce translates the Latinized form of Aristotle's comment: *Nihil est in intellectu quod non prius fuerit in sensu*. In *The Essential Peirce: Selected Philosophical Writings*, ed. N. Houser and C. Kloesel. (Bloomington: Indiana University Press, 1992), 2:227.

39. Carl Hausman, *Charles S. Peirce's Evolutionary Philosophy* (Cambridge: Cambridge University Press, 1993), 115.

40. CP 5:41.

41. CP 1:286.

42. CP 7:219.

43. CP 6:204.

44. CP 2:77 (emphasis added).

45. CP 7:392.

46. MS L75.

47. CP 4:46.

48. CP 4:288.

49. CP 4:299.

50. Peirce supplies a particularly good example of abduction: given that there is a bag of white beans beside me, and given that I have some white beans in my hand, abduction supplies the hypothesis that the beans in my hand came from the bag.

51. For an excellent and detailed exposition of the semiotics of diagrams, see Fredrick Stiernfelt, "Diagrams as Centerpiece of a Peircean Epistemology," *Transactions of the Charles S. Peirce Society* 36, no. 3 (2000): 357–384.

52. Ibid., 365.

53. NEM 59.

54. CP 2:216.

55. Stiernfelt, "Diagrams as a Centerpiece of a Peircean Epistemology," 370.

56. CP 5:436.

57. Henry Wang, "Rethinking the Validity and Significance of Final Causation," *Transactions of the Charles S. Peirce Society* 41, no. 3 (2005): 616.

58. CP 5:514.

59. Beverly Kent, "The Interconnectedness of Peirce's Diagrammatic Thought," in *Studies in the Logic of Charles Sanders Peirce* (Bloomington: Indiana University Press, 1997), 445.

6. ONTOLOGY AND IMAGINATION:
PEIRCE ON NECESSITY AND AGENCY

1. Apel notes that Peirce develops this metaphysics in response to Josiah Royce's *Religious Aspects of Philosophy*, in which one finds Royce's early version of Absolute Idealism. Apel goes on to explain how Peirce's logic and semiotic seeks to correct Royce's system. See Karl-Otto Apel, *From Pragmatism to Pragmaticism*, trans. J. Krois (Amherst: University of Massachusetts Press, 1967), 134–145.

2. CP 6:102.

3. CP 6:59.

4. CP 6:59.

5. MS 954.

6. MS 954.

7. CP 6:60.

8. CP 5:58.

9. Stuart Kauffman, *Investigations* (New York: Oxford University Press, 2000), 8–9; see also *The Origins of Order* (Oxford: Oxford University Press, 1993).

10. MS 292.

11. Carl Hausman, *Charles S. Peirce's Evolutionary Philosophy* (Cambridge: Cambridge University Press, 1993), 172.

12. John Dewey, *A Common Faith*, in *John Dewey: The Late Works: 1925–1953*, ed. J. Boydston (Carbondale, Ill.: SIU Press, 1981–1990), 9:36.

13. Ibid., 9:38.

14. K. T. Fann, *Peirce's Theory of Abduction* (The Hague: Martinus Nijhoff, 1970), 32.

15. Vincent Potter, *Charles Sanders Peirce: On Norms and Ideals* (Amherst: University of Massachusetts Press, 1967), 135. See also Sfendoni-Mentzou's account of continuity and the laws of nature, in which he describes thirdness as such a law: Demetra Sfendoni-Mentzou, "Peirce on Continuity and Laws of Nature," *Transactions of the Charles S. Peirce Society* 33, no. 3 (1997): 649.

16. CP 1:385.

17. Hausman, *Charles S. Peirce's Evolutionary Philosophy*, 168.

18. Here it is tempting to state that Peirce is after the "source" of the imagination, but it seems more faithful to his intent to use the word "principle." Peirce, like many figures from the continental tradition, are interested in interrogating the abductive lead (*Sprung*) that is spoken with the principle (*Satz*) of natural being. As Martin Heidegger later writes, "*Der Sprung ist der Satz aus dem Grundsatz vom Grunde in dem Sagen des Satzes.*"

19. CP 6:102.

20. CP 6:102.

21. MS 954.

22. MS 954.

23. CP 6:288. For a detailed description of the tension between tychism and necessity, see Victor Coscelcuello, "Peirce on Tychism and Determinism," *Transactions of the Charles S. Peirce Society* 38, no. 4 (1992): 741–756.

24. CP 6:288.

25. Maynard Solomón, *Mozart: A Life* (New York: Harper Collins, 1995), 312.

26. This sentiment is expressed repeatedly by the contributors to *Studies of the Logic of Charles Sanders Peirce*, ed. N. Houser and D. Roberts (Bloomington: Indiana University Press, 1997). Robert Burch, for example, criticizes Anderson's *Strands of System: The Philosophy of Charles Sanders Peirce* on the grounds that it concentrates too heavily on ontological-metaphysical aspects of Peirce's thought (embodied in articles such as "Evolutionary Love") as opposed to the "technical-scientific-mathematical-logical side of Peirce's thought." See Robert Burch, "*Strands of System: The Philosophy of Charles Sanders Peirce* by Douglas Anderson," *Philosophy and Literature* 19, no. 2 (1995): 384–385.

27. The translation is mine. The original passage reads: "*Gleich wie keine Vollkommenheit einzeln existieren kann, sondern nur diesen Namen in einer*

gewisen Relation auf einen allgemeinen Zweck verdient, so kann keine denkende Seele sich in sich selbst zurükziehen und mit sich begnügen. . . . Der ewige innere Hang, in das Nebengeschöpf überzugehen, oder dasselbe in sich hineinzuschlingen, es anzureissen ist Liebe." Friedrich Schiller, *Schillers werke: nationalausgabe*, ed. Julius Petersen et al. (Wiemar, 1943), 23:79–80 (emphasis added).

28. Douglas Anderson, *Philosophy Americana: Making Philosophy at Home in American Culture* (New York: Fordham University Press, 2006), 171.

29. John 1:1–3, 14.

30. Charles Peirce, "Evolutionary Love." in *The Essential Peirce*, vol. 1: *Selected Philosophical Writings*, ed. N. Houser and C. Kloesel (Bloomington: Indiana University Press, 1992), 361.

31. Ibid., 369.

32. C. S. Peirce, *Reasoning and the Logic of Things*, ed. K. Kettner. (Cambridge, Mass.: Harvard University Press, 1992), 161. Peirce's use of this common Christian expression, "God lives and moves and has his being in us" (Acts 17:28), may have been inspired proximately from Anna Cabot Lowell's *Seed Grains for Thought*, in which she cites Nathaniel Culverwel's seventeenth-century *Of the Light of Nature*, in which Culverwel states that God is that in which "Reason lives and moves and has its being." See Nathaniel Culverwel, *Of the Light of Nature*, ed. John Cairns (Edinburgh: Thomas Constable, 1856).

33. Translation provided by Martin Heidegger in his *Pathmarks* (Cambridge: Cambridge University Press, 1998), 361.

34. Peirce, *Reasoning and the Logic of Things*, 160.

35. As Peirce begins to conceive his semiotic, he discusses the way in which any finite mind is capable of evolving into an autonomous agent if it has the following three characteristics: (1) awareness broadly construed, (2) action or the ability to affect change, and (3) the ability to take up habits. It will be necessary to map this explanation of what Vincent Colapietro calls the nature of "conscious interpretive agents" onto a careful description of the natural sciences in order to outline the extent to which nature can be considered purposive. See Vincent Colapietro, *Peirce's Approach to the Self: A Semiotic Perspective on Human Subjectivity* (Albany, N.Y.: SUNY Press, 1989), 112.

7. THE EVOLUTION OF THE IMAGINATION

1. Parts of this chapter have appeared in John Kaag, "The Neural Dynamics of the Imagination," *Phenomenology and the Cognitive Sciences* 7, no. 4 (2008): 183–204.

2. George Lakoff and Mark Johnson, *Metaphors We Live By* (Chicago: University of Chicago Press, 1980).

3. George Lakoff and Mark Johnson, *Philosophy in the Flesh: The Embodied Mind and Its Challenge to Western Thought* (New York: Harper Collins, 1999), 45.

4. George Lakoff and Mark Johnson, *Philosophy in the Flesh: The Embodied Mind and Its Challenge to Western Thought* (New York: Harper Collins, 1999), 48. See also the discussion of prototype effects in ibid.

5. Ibid., 541.

6. Donald Dryden, "William James and Susanne Langer: Art and the Dynamics of the Stream of Consciousness," *Journal of Speculative Philosophy* 21, no. 4 (2004): 238.

7. Lakoff and Johnson, *Philosophy in the Flesh*, 25–28; See also George Lakoff, *Women, Fire, and Dangerous Things: What Categories Reveal about the Mind* (Chicago: University of Chicago Press, 1987), 440–444. Gibbs and Colston concluded that these embodied patterns are established early in child development and are stable across cultures. Raymond Gibbs and Herbert Colston, "The Cognitive Psychological Realities of Image Schemas and Their Transformations," *Cognitive Linguistics* 6, no. 4 (2005): 347; Sinha elaborates on the formation of particular image schemas, suggesting that their formation is attributable in large part to the sociocultural forces at play. Christopher Sinha, "Language, Cultural Context, and the Embodiment of Spatial Cognitions," *Cognitive Linguistics* 11, no. 2 (2003): 14–41.

8. Lakoff and Johnson, *Philosophy in the Flesh*, 18.

9. Mark Johnson, *The Body in the Mind: The Bodily Basis of Meaning, Imagination, and Reason* (Chicago: University of Chicago Press, 1987), 29.

10. A cautionary word should be expressed in regard to the character of this metaphor. Lakoff and Johnson are careful to state that "we will . . . use the 'is' in stating . . . MORE IS UP, but the IS should be viewed as a shorthand for some of the experiences on which the metaphor is based and in terms of which we understand it." Lakoff and Johnson, *Metaphors We Live By*, 20.

11. Ibid., 122.

12. As cited in Johnson, *The Body in the Mind*, 32.

13. Ibid., 33.

14. CJ 3:59.

15. CJ 3:59.

16. Carl Hausman, "Metaphorical Reference and Peirce's Dynamical Object," *Transactions of the Charles S. Peirce Society* 23, no. 3 (1987): 381.

17. Lakoff and Johnson, *Philosophy in the Flesh*, 573.

18. Eleanor Rosch, "Cognitive Reference Points," *Cognitive Psychology* 7 (1996): 532–547.

19. Lakoff, *Women, Fire, and Dangerous Things*, 104–114.

20. Claudia Brugman, *The Story of Over: Polysemy, Semantics, and the Structure of the Lexicon* (New York: Garland, 1988).

21. Stuart Kauffman, *Investigations* (Oxford: Oxford University Press, 2000), 135.

22. MS 349.

23. The term "the embodied mind" took hold in the biological sciences in the work of Francisco Varela. During the same period, Varela also helped articulate the concept of "autopoeisis," which will be addressed in the section of the project concerning biology and complexity. See Francisco Varela, *The Principles of Biological Autonomy* (New York: Elsevier North Holland, 1979).

24. Tim Rohrer, "Image Schemata and the Brain," in *Perception to Meaning: Image Schemas in Cognitive Linguistics*, ed. B. Hampe and J. Grady (Berlin: Mouton de Gruyter, 2006), 165.

25. Mark Johnson, *The Meaning of the Body: Aesthetics of Human Understanding* (Chicago: University of Chicago Press, 2007), 45–47.

26. Susan Rose, "Cross-Modal Abilities in Human Infants," *Handbook of Infant Development*, ed. J. Osofsky (New York: Wiley, 1987), 318–362; Arthur Glenberg, "Grounding Language in Action," *Psychonomic Bulletin and Review* 9 (2002): 558–565; Olaf Hauk, "Somatotopic Representations of Action Words in Human Motor and Premotor Cortex," *Neuron* 41 (2004): 301–307; Evelyn Kohler et al., "Hearing Sounds, Understanding Actions: Action Representations in Mirror Neurons," *Science* 297 (2002): 846–848.

27. Gerald Edelman, *Neural Darwinism* (New York: Basic Books, 1987). Edelman goes on to describe the particular mechanisms that grant the possibility of the development of secondary functional repertoires in his concept of "reentry," which stands apart from neural "feedback." This distinction and the imaginative character of reentry will be addressed in the discussion of the organic/molecular basis of the imagination. See Gerald Edelman and Giulio Tononi, *A Universe of Consciousness: How Matter Becomes Imagination* (New York: Basic Books, 2000), 48.

28. Donald Hebb, *The Organization of Behavior: A Neuropsychological Theory* (New York: Wiley, 1949), 45.

29. Elie Bienenstock et al., "Theory for the Development of Neuronal Selectivity: Orientation Specificity and Binocular Interaction in the Visual Cortex," *Journal of Neuroscience* 2, no. 1 (1982): 32–48.

30. John Byrne, *Neural Models of Plasticity* (San Diego, Calif.: Academic Press, 1989).

31. Jeffrey Schwartz and Sharon Begley, *The Mind and the Brain: Neuroplasticity and the Power of Mental Force* (New York: Regent, 2002), 118.

32. CP 4:39.

33. William James, *The Principles of Psychology* (New York: Dover, 1950), 262.

34. Rohrer, "Image Schemata and the Brain," 19.

35. Charles Gallistel, *The Organization of Learning* (Cambridge, Mass.: MIT Press, 1990).

36. Gerald Edelman, *The Remembered Present: A Biological Theory of Consciousness* (New York: Basic Books, 1989), 50.

37. Ibid.

38. Ibid., 48.

39. Ibid., 50.

40. Ibid., 22.

41. Ibid.

42. Terry Allard et al., "Reorganization of Somatosensory Area 3b Representations in Adult Owl Monkeys after Digital Syndactyly," *Journal of Neurophysiology* 66 (1991): 1048–1058.

43. Jeffrey Fox, "The Brain's Dynamic Way of Keeping in Touch," *Science* 225, no. 4664 (1984): 820.

44. Jerome Feldman, *From Molecule to Metaphor* (Cambridge, Mass.: MIT Press, 2006), 105.

45. Ibid., 331.

46. M. Umilta et al., "I Know What You Are Doing: A Neurophysiological Study," *Neuron* 31 (2001): 155–165.

47. Ibid.

48. V. Gallese, "Action Recognition in the Premotor Cortex," *Brain* 119 (1996): 593–609.

49. E. Kohler, "Hearing Sounds, Understanding Actions: Action Representation in Mirror Neurons," *Science* 297 (2002): 846.

50. M. Umilta et al., "I Know What You Are Doing," 155.

8. EMERGENCE, COMPLEXITY, AND CREATIVITY

1. Parts of this chapter have appeared in John Kaag, "The Neural Dynamics of the Imagination," *Phenomenology and the Cognitive Sciences* 7, no. 4 (2008): 183–204.

2. Donald Tucker, *Mind from Body: Neural Structures of Experience* (New York: Oxford University Press, 2007), 168.

3. CP 4:45.

4. Tucker, *Mind from Body*, 13.

5. This point is made by Mark Johnson, "Dewey's Zen: The 'Oh' of Wonder," discussion paper presented at the Society for the Advancement of American Philosophy, 2007.

6. Tucker, *Mind from Body*, 169. See also Benjamin Libet, "Brain Stimulation in the Study of Neuronal Functions for Conscious Sensory Experiences." *Human Neurobiology* 1: 235–242.

7. Tucker, *Mind from Body*, 169.

8. Friedrich Schiller, *Samtliche Werke*, ed. Gerhard Fricke et al. (Munich, 1960), 633.

9. Tucker, *Mind from Body*, 166.

10. The child-candle example is drawn from William James, *The Principles of Psychology* (New York: Henry Holt, 1916), 2:34–35.

11. John Dewey, "The Reflex Arc Conception in Psychology," *Psychological Review* 3 (1896): 357.

12. The concept of reentry has been addressed by many researchers for nearly two decades. See O. Sporns et al., "Modeling Perceptual Grouping and Figure-Ground Segregation by Means of Active Reentrant Connections," *Proceedings of the National Academy of Sciences of the United States of America* 88 (1991): 129–133. See also Gerald Edelman, "Neural Dynamics in a Model of the Thalamocortical System, 2: The Role of Neural Synchrony Tested through the Perturbations of Spike Timing," *Cerebral Cortex* 7 (1997): 228–236.

13. Gerald Edelman and J. Gally, "Degeneracy and Complexity in Biological Systems," *Proceedings of the National Academy of Sciences of the United States of America* (November 6, 2001): 13763–13768.

14. Gerald Edelman and Giulio Tononi, *A Universe of Consciousness: How Matter Becomes Imagination* (New York: Basic Books, 2000), 104–105.

15. CP 6:104.

16. Edelman and Tononi, *A Universe of Consciousness*, 49.

17. Gerald Edelman, *Neural Darwinism: The Theory of Neuronal Group Selection* (New York: Basic Books, 1987), 61.

18. Ibid.

19. Gerald Edelman, "Building a Picture of the Brain," *Annals of the New York Academy of Science* 882, no. 1 (1999): 70.

20. Edelman and Tononi, *A Universe of Consciousness*, 64.

21. Ibid., 114.

22. Ibid., 86–87, 97–98. See also G. Tononi et al., "Measures of Degeneracy and Redundancy in Biological Networks," *Proceedings of the National Academy of Sciences of the United States of America* 96 (1995): 3188–3208.

23. Edelman and Gally, "Degeneracy and Complexity in Biological Systems."

24. Edelman and Tononi, *A Universe of Consciousness*, 98.

25. Edelman and Tononi, *A Universe of Consciousness*, 86–87.

26. CP 6:57.

27. CP 6:45.

28. Edelman and Gally, "Degeneracy and Complexity in Biological Systems," 13767.

29. Edelman and Tononi, *A Universe of Consciousness*, 149.

30. For material on the relation between conceptual generalization and the detection of relative similarity and difference, see R. Shepard et al., "The Analysis of Proximities: Multidimensional Scaling with an Unknown Distance Function. II," *Psychometrika* 27 (1962): 219–246; and "Toward a Universal Law of Generalization for Psychological Science," *Science* 237 (1987): 1317–1323.

31. Peter Gardenfors, *Conceptual Spaces: A Geometry of Thought* (Cambridge, Mass.: MIT Press, 2000), 109.

32. Similar conclusions have been reached in the recent computational work of Paul Churchland, in which he examines the similarity across sensory and neural space. Paul Churchland, "Conceptual Similarity across Sensory and Neural Diversity: The Fodor/Lepore Challenge Answered," *Journal of Philosophy* 95 (1998): 5–32.

33. Gardenfors, *Conceptual Spaces*, 123.

34. Ibid., 122.

35. Ibid.

36. Ibid.

37. Ibid., 160. See also P. Langley, *Elements of Machine Learning* (San Francisco: Morgan, 1996), 99–100.

38. CP 2:776

39. John Holland, *Hidden Order: How Adaptation Gives Rise to Complexity* (New York: Perseus, 2002), 56.

40. Note that this description of mental complexity bears a marked similarity to Peirce's description of a continuous system that cannot be accurately described as the discrete sum of any particular set of individual actors.

41. Humberto Maturana, "Autopoeisis: The Organization of the Living," in *Autopoeisis and Cognition* (Dordrecht: Reidel, 1980). Humberto Maturana and Francisco Varela, *The Tree of Knowledge* (Boston: Shambhala, 1987). See also John Mingers, *Self-Producing Systems* (New York: Plenum, 1995).

42. Vittorio Gallese and Thomas Metzinger, "The Emergence of a Shared Action Ontology: Building Blocks for a Theory," *Consciousness and Cognition* 12 (2003): 549–571.

43. Thomas Metzinger, *Being No One: The Self-Model of Subjectivity* (Cambridge, Mass.: MIT Press, 2004), 414.

44. It should be noted that Gallese's intention is to set the stage for future investigation into the biological basis of interpersonal relations, empathy, compassion, etc. He believes that the distinctions he draws between action and movement allow him to explain the emergence of what he considers a uniquely human ability to have a *shared* action ontology. The goal of my project is somewhat different. In reviewing the empirical literature, it comes to accept a "shared action ontology" in another sense of the expression: organic being shares a particular form of action best described as imaginative. See Gallese and Metzinger, "The Emergence of a Shared Action Ontology," 568.

45. CJ 253.

46. See Holland, *Hidden Order*; and Stuart Kauffman, *Investigations* (Oxford: Oxford University Press, 2000).

47. In providing a syllogism and hypothesis ("if . . . then" structure) as examples of rule-based agency, Holland echoes Peirce's suspicion in his 1898 Cambridge lectures that we think of nature as syllogizing. Holland, *Hidden*

Order, 22, 44–91. As a side note, Kauffman, also in line with Peirce's pragmatic tendencies, suggests that it may be frustrating, if not futile, to search for an elemental agent whose irreducible rule set could be easily defined. See Kauffman, *Investigations.*

48. Holland, *Hidden Order*, 11.

49. Ibid., 31.

50. Ibid., 27–30.

9. *BE IMAGINATIVE!* SUGGESTION AND IMPERATIVE

1. John Dewey, *Art as Experience* [1934], in *John Dewey: The Later Works, 1925–1953*, ed. Jo Ann Boydson (Carbondale: Southern Illinois University Press, 1986), 12:277.

2. Colin McGinn, "Can We Solve the Mind-Body Problem?" *Mind* 98 (1989): 355.

3. Douglas R. Anderson and Carl R. Hausman, *Conversations on Peirce* (Bronx, N.Y.: Fordham University Press, 2012), 34.

4. Cited in Raymond Boisvert, *John Dewey: Rethinking Our Time* (Albany, N.Y.: SUNY Press, 1998), 154.

5. "John Dewey to Read Bain," in *The Correspondence of John Dewey, 1871–1952.* Vol. 2. Electronic ed., 1935.03.12 (07781). Charlottesville, Va.: InteLex, 2008.

6. "James Feibleman to John Dewey," in ibid., 1939.02.10 (07950).

7. "John Dewey to James Feibleman," in ibid., 1935.03.12 (07781), 1939.02.16 (07951).

8. Maurice Merleau-Ponty, "Cezanne's Doubt," in *Sense and Non-Sense* (Evanston, Ill.: Northwestern University Press, 1964), 15.

9. MS 45 (now placed in L 299).

10. "To Ferdinand Canning Scott Schiller," William James Papers, Houghton Library, Harvard University, bMs Am 1092.9 (3702).

11. Donald Dryden, "Susanne Langer and William James: Art and the Dynamics of the Stream of Consciousness," *Journal of Speculative Philosophy* 21, no. 4 (2001): 282.

12. Susanne Langer, *Philosophy in a New Key* (Cambridge, Mass.: Harvard University Press, 1957), 46.

13. Henry David Thoreau, *Walden and Other Writings*, ed. J. W. Krutch (New York: Bantam, 1981), 220 (emphasis added).

14. Ibid., 223.

15. Ibid., 224.

16. Ibid., 227.

17. Ibid.

18. Ibid., 228.

19. Ibid., 200.

20. William James, *The Will to Believe and Other Essays in Popular Philoso-phy* (New York: Longmans Green and Co., 1896), 27.

21. Charles Peirce, *The Collected Writings of Charles Sanders Peirce*, ed. M. Fisch and C. Kloesel (Bloomington: Indiana University Press, 1982), 1:xxix.

22. Ibid., 1:45–46.

23. Ibid., 1:xxvii.

Bibliography

Kant on Imagination: Primary Literature

Kant, Immanuel. *The Critique of Judgment* [1790]. Trans. W. Pluhar. Indianapolis: Hackett, 1987.
———. *The Critique of Pure Reason* [1781]. Trans. W. Pluhar. Indianapolis: Hackett, 1996.
———. *Dreams of a Spirit Seer, Illustrated by Dreams of Metaphysics*. Trans. E. Goerwitz. New York: Macmillan, 1900.
———. *Metaphysical Foundations of Natural Science*. Trans. J. Ellington. Indianapolis: Bobbs Merrill, 1970.
———. *Opus Postumum* [1794]. Trans. E. Förster. Cambridge: Cambridge University Press, 1993.

Kant on Imagination: Secondary Literature

Allison, Henry. *Kant's Theory of Taste: A Reading of the* Critique of Judgment. Cambridge: Cambridge University Press, 2001.
———. *Kant's Transcendental Idealism: An Interpretation and Defense*. New Haven, Conn.: Yale University Press, 2004.
Arendt, Hannah. *Lectures on Kant's Political Philosophy*. Ed. Ronald Beiner. Chicago: University of Chicago Press, 1982.
———. *The Life of the Mind*. New York: Harcourt, 1977.
Cohen, T., and P. Guyer, eds. *Essays in Kant's Aesthetics*. Chicago: University of Chicago Press, 1982.
Cornell, Drucilla. "Enlightening and Enlightenment: A Response to John Brenkman." *Critical Inquiry* 26, no. 1 (1999).
Crawford, Donald. *Kant's Aesthetic Theory*. Madison: University of Wisconsin Press, 1974.
———. "Kant's Theory of Creative Imagination." In *Essays on Kant's Aesthetics*, ed. T. Cohen and P. Guyer. Chicago: University of Chicago Press, 1982.
Duve, Thierry de. *Kant After Duchamp*. Cambridge, Mass.: MIT Press, 1996.

Fischer, Kuno. *A Commentary on Kant's* Critik of Pure Reason. London: Longman and Green, 1866.

Förster, E. *Kant's Final Thesis: An Essay on the* Opus Postumum. Cambridge, Mass.: Harvard University Press, 2000.

———. "Kant's *Selbstsetzungslehre.*" In *Kant's Transcendental Deductions,* ed. E. Förster. Stanford, Calif.: Stanford University Press, 1989.

———. "Kant's Third *Critique* and the *Opus Postumum.*" *Graduate Faculty Philosophy Journal* 16 (1993): 345–358.

Gadamer, Hans-Georg. *Truth and Method.* New York: Continuum, 1995.

Gasché, Rodolphe. *The Idea of Form: Rethinking Kant's Aesthetics.* Stanford, Calif.: Stanford University Press, 2003.

———. "Leaps of Imagination." In *The Path of Archaic Thinking: Unfolding the Work of John Sallis,* ed. K. Maly. Albany, N.Y.: SUNY, 1995.

Gregor, M. "Aesthetic Form and Sensory Content in the *Critique of Judgment*: Can Kant's Critique of Aesthetic Judgment Provide a Philosophical Basis for Modern Formalism?" In *The Philosophy of Immanuel Kant,* ed. R. Kennington. Washington, D.C., 2000.

Guyer, Paul. *Kant and the Claims of Taste.* Cambridge, Mass.: Harvard University Press, 1979.

———. *Kant and the Experience of Freedom: Essays on Aesthetics and Morality.* New York: Cambridge University Press, 1993.

Heidegger, M. *Pathmarks.* Ed. W. McNeill. Cambridge: Cambridge University Press. 1998.

———. *Phenomenological Interpretation of Kant's* Critique of Pure Reason. Trans. P. Emad. Bloomington: Indiana University Press, 1997.

Makkreel, R. *Imagination and Interpretation in Kant: The Hermeneutical Import of the* Critique of Judgment. Chicago: University of Chicago Press, 1990.

Sallis, J. *Delimitations: Phenomenology and the End of Metaphysics.* Bloomington: Indiana University Press, 1995.

———. *Force of Imagination: The Sense of the Elemental.* Bloomington: Indiana University Press, 2000.

———. *The Gathering of Reason.* Athens: Ohio University Press, 1980.

———. *Shades of Painting at the Limit.* Bloomington: Indiana University Press, 1998.

Schaper, Eva. "Taste, Sublimity, and Genius: The Aesthetics of Nature and Art." In *The Cambridge Companion to Kant,* ed. P. Guyer. Chicago: University of Chicago Press, 1982.

Schiller, Friedrich. *Samliche Werke.* Ed. Gerhard Fricke et al. Munich, 1960.

Sobel, J. E. "Arguing, Accepting, and Preserving Design." In *Essays in Kant's Aesthetics,* ed. T. Cohen and P. Guyer. Cambridge: Cambridge University Press, 1982.

Zammito, J. *The Genesis of Kant's* Critique of Judgment. Chicago: University of Chicago Press, 1992.

Peirce on Imagination: Primary Literature

The Writings of C. S. Peirce: A Chronological Edition. 6 vols. Ed. C. Hartshorne and P. Weiss. Cambridge, Mass.: Harvard University Press.
———. "Private Thoughts—Principally on the Conduct of Life" (1853).
———. "Analysis of Genius" (1859).
———. "On the Doctrine of Immediate Perception" (1864).
———. "Letter, Peirce to Francis E. Abbot" (1865).
——— "Harvard Lecture on Kant" (1865).
———. "Harvard Lecture VIII" (1865) [Peirce's rejoinder to the lecture on Kant].
———. "Searching for the Categories" (1866).
———. "Questions Concerning Certain Faculties" (1868).
———. "Henry James's 'The Secret of Swedenborg'" (1870).
———. "On Time and Thought" (1873).
———. "Deduction, Induction, and Hypothesis" (1878).
———. "Logic. Chapter I. Of Thinking as Cerebration" (1879).
———. "The Beginnings of a Logic Book" (1883).
———. "Notes on the Categories" (1885).
———. "One, Two, Three: Fundamental Categories of Thought and of Nature" (1885).
———. "Kant's Introduction to Logic" (1885).
———. "One, Two, Three: Kantian Categories" (1886).
———. "First, Second, Third" (1886).
———. "Circular for the Course on the Art of Reasoning" (1887).
———. "Letter to Noble on the Nature of Reasoning" (1887).
The Essential Peirce: Selected Philosophical Writings. 2 vols. Ed. Peirce Edition Project. Bloomington: Indiana University Press, 1998.
———. "Fraser's The Work of George Berkley" (1871).
———. "The Doctrine of Chances" (1878).
———. "The Order of Nature" (1878).
———. "Design and Chance" (1883/1884).
———. "A Guess at the Riddle" (incomplete book, 1887/1888).
———. "The Architecture of Theories" (1891).
———. "Laws of Nature" (1891).
———. "The Doctrine of Necessity Examined" (1892).
———. " The Law of Mind" (1892).
———. "Man's Glassy Essence" (1892).
———. "Evolutionary Love" (1893).

———. "Of Reasoning in General" (1895).

———. "The Maxim of Pragmatism" (1903).

———. "On Phenomenology" (1903).

———. "The Seven Systems of Metaphysics" (1903).

———. "The Nature of Meaning" (1903).

———. "Pragmatism as the Logic of Abduction" (1903).

———. "What Pragmatism Is" (1904).

———. "Pragmatism" (1907).

———. "Neglected Argument for the Reality of God" (1908).

———. "An Essay toward Improving Our Reasoning in Security and Liberty" (1913).

Peirce and the Imagination: General and Secondary Literature

Abbot, Francis Ellington. *The Syllogistic Philosophy*. Boston: Little Brown and Company, 1906.

Ales Bello, Angela. "Peirce and Husserl: Abduction, Apperception, and Aesthetics." In *Peirce and Value Theory*, ed. H. Parret. Amsterdam: Benjamins, 1994.

Alexander, T. *John Dewey's Theory of Art, Experience, and Nature: The Horizon of Feeling*. Albany, N.Y.: SUNY Press, 1987.

———. "Pragmatic Imagination." *Transactions of the Charles S. Peirce Society* 26, no. 3 (1990).

Anderson, D. *Creativity and the Philosophy of C. S. Peirce*. Boston: Martinus Nijhoff, 1987.

———. "Peirce and Metaphor." *Transactions of the Charles S. Peirce Society* 20 (1984): 453–468.

———. "Peirce's Common Sense Marriage of Religion and Science." In *The Cambridge Companion to Peirce*, ed. C. Misak. Cambridge: Cambridge University Press, 2004.

———. *Strands of System: The Philosophy of Charles Peirce*. West Lafayette, Ind.: Purdue University Press, 1995.

Apel, K. "From Kant to Peirce: The Semiotical Transformation of Transcendental Logic." In *Proceedings of the Third International Kant Congress*, ed. Lewis White Beck, 90–104. Dordrecht: D. Reidle, 1972.

———. *From Pragmatism to Pragmaticism*. Trans. J. Krois. Amherst: University of Massachusetts Press, 1981.

Ayer, A. J. *The Origins of Pragmatism: Studies in the Philosophy of Charles Sanders Peirce and William James*. London: Macmillan, 1968.

Barnouw, Jeffrey. "'Aesthetic' for Schiller and Peirce: A Neglected Origin of Pragmatism." *Journal of the History of Ideas* 49 (1988): 607–632.

——. "The Place of Peirce's Esthetic in His Thought and in the Tradition of Aesthetics." In *Peirce and Value Theory*, ed. H. Parret. Amsterdam: Benjamins, 1994.

Bedell, Gary. "Has Peirce Refuted Egoism?" *Transactions of the Charles S. Peirce Society* 16 (1980): 255–275.

Beil, R., and K. Ketner. "Peirce, Clifford, and Quantum Theory." *International Journal of Theoretical Physics* 42, no. 9 (2003).

Boersema, David. "On Peirce on Man as Language." *Kinesis* 13 (1984): 65–76.

——. "Peirce on Explanation." *Journal of Speculative Philosophy* 17, no. 3 (2003): 224–236.

Boisvert, R. *John Dewey: Rethinking Our Time*. Albany, N.Y.: SUNY Press, 1998.

Brown, W. M. "The Economy of Peirce's Abduction." *Transactions of the Charles S. Peirce Society* 19, no. 4 (1983).

Bruning, R., and G. Lohmann. "Charles S. Peirce on Creative Metaphor: A Case Study on the Conveyor Belt Metaphor in Oceanography." *Foundations of Science* 4, no. 4 (1999): 389–403.

Buchler, Justus. *Charles Peirce's Empiricism*. London: Keagan, 1939.

Burks, Arthur. "Logic, Learning, and Creativity in Evolution." In *Studies in the Logic of Charles Sanders Peirce*, ed. N. Houser. Bloomington: Indiana University Press, 1997.

Calabrese, Omar. "Some Reflections on Peirce's Aesthetics from a Structuralist Point of View." In *Peirce and Value Theory*, ed. H. Parret. Amsterdam: Benjamins, 1994.

Chauvire, Christiane. "Schematisme et analyticite chez C S Peirce." *Archives de Philosophie* 50 (1987).

Cheah, P. "Human Freedom and the Technic of Nature: Culture and Organic Life in Kant's Third *Critique*." *Differences: A Journal of Feminist Cultural Studies* 14 (2003).

Code, M. "Interpreting the 'Raw Universe' Meaning and Metaphysical Imaginaries." *Transactions of the Charles S. Peirce Society* 35, no. 4 (1999).

Colapietro, V. *Peirce's Approach to the Self*. Albany, N.Y.: SUNY Press, 1989.

Cosculluela, Victor. "Peirce on Tychism and Determinism." *Transactions of the Charles S. Peirce Society* 28, no. 4 (1992): 741–755.

Cozzo, C. "Epistemic Truth and Excluded Middle." *Theoria* 64, nos. 2/3 (1998): 243–282.

Davenport, William. "Peirce on Evolution by Revolution." In *Frontiers in American Philosophy*, ed. R. W. Burch, 2:293–301. College Station: Texas A&M University Press, 1996.

De Tienne, A. "Peirce's Early Method of Finding the Categories." *Transactions of the Charles S. Peirce Society* 25 (1989): 385–407.

Decker, K. "Ground, Relation, Representation: Kantianism and the Early Peirce." *Transactions of the Charles S. Peirce Society* 37, no. 2 (2001): 179–206.

Dewey, John. *Art as Experience.* Vol. 12 of *John Dewey: The Later Works, 1925–1953.* Ed. Jo Ann Boydson. Carbondale: Southern Illinois University Press, 1987.

———. "The Reflex Arc Conception in Psychology." *Psychological Review* 3 (1896).

Dryden, Donald. "Susanne Langer and William James: Art and the Dynamics of the Stream of Consciousness." *Journal of Speculative Philosophy* 21, no. 4 (2001).

Esposito, J. "The Development of Peirce's Categories." *Transactions of the Charles S. Peirce Society* 15 (1979).

———. *Evolutionary Metaphysics.* Athens: University of Ohio Press, 1980.

———. "Peirce and the Philosophy of History." *Transactions of the Charles S. Peirce Society* 19 (1983): 155–166.

Fairbanks, Matthew. "Peirce on Man as a Language: A Textual Interpretation." *Transactions of the Charles S. Peirce Society* 12 (1976): 18–32.

Fann, K. T. *Peirce's Theory of Abduction.* The Hague: Martinus Nijhoff, 1970.

Frankfurt, Harold. "Peirce's Notion of Abduction." *Journal of Philosophy* 55 (1958).

Garrison, James. "The Logic, Ethics, and Aesthetics of Geometrical Construction." In *Peirce and Value Theory,* ed. H. Parret. Amsterdam: Benjamins, 1994.

Harman, Gilbert. "Inference to Best Explanation." *Philosophical Review* 74 (1965).

Hatten, Robert. "A Peircean Perspective on the Growth of Markedness and Musical Meaning." In *Peirce and Value Theory,* ed. H. Parret. Amsterdam: Benjamins, 1994.

Hausman, C. "Bergson, Peirce, and Reflective Intuition." *Process Studies* 28, nos. 3/4 (1999): 289–300.

———. "Charles Peirce and the Origin of Interpretation." In *The Rule of Reason: The Philosophy of Charles Sanders Peirce,* ed. P. Forster, 185–200. Toronto: University of Toronto Press, 1997.

———. *Charles S. Peirce's Evolutionary Philosophy.* Cambridge: Cambridge University Press, 1993.

———. "Freedom, Indeterminism, and Necessity in the Origination of Novelty." *Southern Journal of Philosophy* 9: 170–189.

———. "Insight in the Arts." *Journal of Aesthetics and Art Criticism* (Winter 1986).

Healy, Kevin. "Peirce, Community, and Belief." *Prospero* 2, no. 2 (1996): 56–61.

Hickman, L. *John Dewey's Pragmatic Technology.* Bloomington: Indiana University Press, 1990.

Hintikka, Jaakko. "What Is Abduction? The Fundamental Problem of Contemporary Epistemology." *Transactions of the Charles S. Peirce Society* 34, no. 3 (1998).

Hookway, C. *Peirce.* London: Routledge, 1985.

———. "Truth, Reality, and Convergence." In *The Cambridge Companion to Peirce,* ed. C. Misak, 127–149. Cambridge: Cambridge University Press, 2004.

Josephson, John, et al. *Abductive Inference: Computation, Philosophy, and Technology.* Cambridge: Cambridge University Press, 1994.

Kaag, John. "Continuity and Inheritance: Kant's *Critique of Judgment* and the Philosophy of C. S. Peirce." *Transactions of the Charles S. Peirce Society* 41, no. 3 (2005): 515–540.

Kapitan, T. "In What Way Is Abductive Inference Creative?" *Transactions of the Charles Peirce Society* 26 (1990): 499–512.

———. "Peirce and the Autonomy of Abductive Reasoning." *Erkenntnis* 37, no. 1 (1992): 1–26.

———. "Peirce and the Structure of Abductive Inference." In *Studies in the Logic of Charles Sanders Peirce,* ed. N. Houser et al. Bloomington: Indiana University Press, 1997.

Kemp Pritchard, Ilona. "Peirce on Individuation." *Transactions of the Charles S. Peirce Society* 14 (1978): 83–100.

Kent, Beverly. *Charles S. Peirce: Logic and the Classifications of the Sciences.* Kingston, Ont.: McGill University, 1987.

Ketner, K. *The Elements of Logic: An Introduction to Peirce's Existential Graphs.* Lubbock, Tex.: Arisbe, 1996.

———. *His Glassy Essence: An Autobiography of Charles Sanders Peirce.* Nashville, Tenn.: Vanderbilt University Press, 1998.

Kevelson, Roberta. "The Mediating Role of Esthetics in Charles S. Peirce's Semiotics." In *Peirce and Value Theory,* ed. H. Parret. Amsterdam: Benjamins, 1994.

Kuklick, Bruce. *A History of Philosophy in America, 1720–2000.* Oxford: Oxford University Press, 2001.

Johnstone, Henry. "Charles Peirce: Philosopher of Science and Common Sense." *Hermathena* 96 (1962): 3–15.

Langer, Susanne. *Philosophy in a New Key: A Study in the Symbolism of Reason, Rite, and Art.* Cambridge, Mass.: Harvard University Press. 1957.

Levi, Albert. "Peirce and Painting." *Philosophy and Phenomenological Research* 23 (1962): 23–36.

Magada-Ward, Mary. "'As Parts of One Esthetic Total:' Inference, Imagery, and Self-Knowledge in the Later Peirce." *Journal of Speculative Philosophy.* 17(3): 216–223, 2003.

Magnani, Lorenzo. *Abduction, Reason, and Science: Processes of Discovery and Explanation.* New York: Kluwer, 2001.

Marcio, Jaime. "Thought Is Essentially an Action: Peirce and Rorty on Normal and Abnormal Discourse." *Inquiry* 21, no. 1 (2001): 33–42.

Marsoobian, Armen. "Art and Interpretation: Peirce and Buchler on Aesthetic Meaning." In *Peirce and Value Theory*, ed. H. Parret. Amsterdam: Benjamins, 1994.

Merrell, Floyd. *Peirce, Signs, and Meaning*. Toronto: University of Toronto Press, 1997.

Michael, E. "Peirce's Adaptation of Kant's Definition of Logic in the Early Manuscripts." *Transactions of the Charles Peirce Society* 14 (1978): 176–184.

Murphey, M. *The Development of Peirce's Philosophy*. Cambridge, Mass.: Harvard University Press, 1961.

———. "Kant's Children: The Cambridge Pragmatists." *Transactions of the Charles S. Peirce Society* 4 (1968): 3–35.

Noble, B. "Peirce's Definition of Continuity and the Concept of Possibility." *Transactions of the Charles S. Peirce Society* 25 (1989): 149–174.

Oehler, Klaus. "A Response to Habermas." In *Peirce and Contemporary Thought: Philosophical Inquiries*, ed. K. Ketner. Bronx, N.Y.: Fordham University Press: 1995.

Pape, Helmut. "Love's Power and the Causality of Mind: C. S. Peirce on the Place of Mind and Culture in Evolution." *Transactions of the Charles S. Peirce Society* 33, no. 1 (1997): 59–90.

Parker, Kelly. *The Continuity of Peirce's Thought*. Nashville, Tenn.: Vanderbilt University Press, 1997.

Popkin, Richard. "Early Influences on Peirce: A Letter to Samuel Barnett." *Journal of the History of Philosophy* 31, no. 4 (1993): 607–621.

Portis Winner, Irene. "Peirce, Saussure, and Jakobson's Aesthetic Function." In *Peirce and Value Theory*, ed. H. Parret. Amsterdam: Benjamins, 1994.

Potter, V. *Charles S. Peirce: Norms and Ideals*. Bronx, N.Y.: Fordham University Press, 1997.

———. "Peirce's Definitions of Continuity." *Transactions of the Charles S. Peirce Society* 13 (1977): 20–24.

Robinson, Andrew. "Continuity, Naturalism, and Contingency: A Theology of Evolution Drawing on the Semiotics of C. S. Peirce and Trinitarian Thought." *Zygon* 39, no. 1 (2004): 111–136.

Rosenthal, C. "A Pragmatic Appropriation of Kant: Lewis and Peirce." *Transactions of the Charles S. Peirce Society* 38, nos. 1/2 (2002): 253–266.

Rosenthal, S. "C. S. Peirce: Pragmatism, Semiotic Structure, and Lived Perceptual Experience." *Journal of the History of Philosophy* 17 (1979): 285–290.

———. *Charles S. Peirce's Pragmatic Pluralism*. Albany, N.Y.: SUNY Press, 1994.

———. "Peirce's Pragmatic Account of Perception: Issues and Implications." In *The Cambridge Companion to Peirce*, ed. C. Misak, 193–213. Cambridge: Cambridge University Press, 2004.

Salabert, Pere. "Aesthetic Experience in Charles S Peirce: The Threshold." In *Peirce and Value Theory*, ed. H. Parret. Amsterdam: Benjamins, 1994.

Schum, David. "Evidence Marshalling for Imaginative Fact Investigation." *Artificial Intelligence and Law* 9 (2001).

Scott, Frederick. "Peirce and Schiller and Their Correspondence." *Journal of the History of Philosophy* 11 (1973): 363–386.

Seigfried, C. *Chaos and Context: A Study of William James*. Athens: Ohio University Press, 1978.

———. *William James' Radical Reconstruction of Philosophy*. Albany, N.Y.: SUNY Press, 1990.

Sfendoni Mentzou, Demetra. "Peirce on Continuity and Laws of Nature." *Transactions of the Charles S. Peirce Society* 33, no. 3 (1997): 646–678.

Shin, S. "Kant's Syntheticity Revisited by Peirce." *Synthese* 113, no. 1 (1997): 1–41.

Smith, C. "The Aesthetics of Charles S. Peirce." *Journal of Aesthetics and Art Criticism* 31 (1972): 21–29.

Sorrell, Kory. "Peirce and a Pragmatic Reconception of Substance." *Transactions of the Charles S. Peirce Society* 37, no. 2 (2001): 257–295.

Sowa, John. "Matching Logical Structure to Linguistic Structure." In *Studies in the Logic of Charles Sanders Peirce*, ed. N. Houser, 418–444. Bloomington: Indiana University Press, 1997.

Stephens, Lynn. "Noumenal Qualia: C. S. Peirce on Our Epistemic Access to Feelings." *Transactions of the Charles S. Peirce Society* 21 (1985): 95–108.

Tejera, Victorino. "Has Habermas Understood Peirce?" *Transactions of the Charles S. Peirce Society* 32, no. 1 (1996): 107–125.

———. "The Primacy of the Aesthetic in Peirce, and Classic American Philosophy." In *Peirce and Value Theory*, ed. H. Parret. Amsterdam: Benjamins, 1994.

Tiercelin, Claudine. "Peirce on Norms, Evolution, and Knowledge." *Transactions of the Charles S. Peirce Society* 33, no. 1 (1997): 35–58.

Thompson, Manley. *The Pragmatic Philosophy of Charles S. Peirce*. Chicago: University of Chicago Press, 1953.

Thoreau, Henry David. *Walden*. Ed. J. Krutch. New York: Bantam, 1981.

Trammell, Richard. "Charles Sanders Peirce and Henry James the Elder." *Transactions of the Charles S. Peirce Society* 9 (1973): 202–217.

Walton, Douglas. *Abductive Reasoning*. Tuscaloosa: University of Alabama Press, 2004.

Wang, Henry. "Rethinking the Validity and Significance of Final Causation." *Transactions of the Charles S. Peirce Society* 41, no. 3 (2005).

Weiner, Phillip. "Peirce's Evolutionary Interpretation of the History of Science." In *Studies in the Philosophy of Charles S. Peirce*, ed. Phillip Wiener and Fredric Young. Cambridge, Mass.: Harvard University Press, 1952.

Wills, Fredrick. *Beyond Deduction*. New York: Routledge, 1988.

Wilshire, B. *William James and Phenomenology: A Study of the* Principles of Psychology. Bloomington: Indiana University Press, 1968.

Zeman, Jay. "Peirce on the Indeterminate and on the Object: Initial Reflections." *Grazer Philosophische Studien* 32 (1988): 37–49.

Imagination and Cognitive Science

Allard, Terry, et al. "Reorganization of Somatosensory Area 3b Representations in Adult Owl Monkeys after Digital Syndactyly." *Journal of Neurophysiology* 66 (1991): 1048–1058.

Arnheim, Rudolf. *Toward a Psychology of Art: Collected Essays*. Berkeley: University of California Press, 1966.

Barsalou, L. "Ad hoc Categories." *Memory and Cognition* 10 (1983): 82–93.

———. "Flexibility, Structure, and Linguistic Vagary in Concepts: Manifestations of a Compositional System of Perceptual Symbols." In *Theories of Memory*. Hilldale: Lawrence and Associates, 1992.

———. "The Instability of Graded Structure: Implications for the Nature of Concepts." In *Concepts and Conceptual Development: The Ecological and Intellectual Factors of Categorization*. Cambridge: Cambridge University Press, 1987.

———. "Mundane Creativity in Perceptual Symbol Systems." In *Creative Thought: An Investigation of Conceptual Structure and Processes*. Washington, D.C.: American Psychological Association, 1997.

Black, M. "More about Metaphor." In *Metaphor and Thought*. Cambridge: Cambridge University Press, 2002.

Brink, I. "Metaphor, Similarity, and Semantic Fields." *In Understanding the Arts: Contemporary Aesthetics*. Lund: Lund University, 1992.

Brostrom, S. *The Role of Metaphor in Cognitive Semantics*. Lund: Lund University Cognitive Studies, 1994.

Brugman, Claudia. *The Story of Over*. Bloomington: Indiana Linguistic Club, 1981.

Capra, Fritjof. *The Web of Life: A New Scientific Understanding of Living Systems*. New York: Anchor, 1996.

Casti, John. *Complexification: Explaining a Paradoxical World through the Science of Surprise*. New York: HarperCollins, 1994.

Churchland, P. "Conceptual Similarity across Sensory and Neural Diversity: The Fodor/Lepore Challenge Answered." *Journal of Philosophy* 95: 5–32.

Colston, H. "The Cognitive Psychological Realities of Image Schemas and Their Transformations." *Cognitive Linguistics* 6, no. 4: 347.

Damasio, Antonio. *Descartes' Error: Emotions, Reason, and the Human Brain.* New York: Putnam, 1994.

———. *The Feeling of What Happens: The Body and Emotion in the Formings of Consciousness.* New York: Harcourt, 1999.

———. *Looking for Spinoza: Joy, Sorrow, and the Human Brain.* Orlando, Fla.: Harcourt, 2003.

———. "Some Notes on Brain, Imagination and Creativity." In *The Origins of Creativity*, ed. K. Pfenninger and V. Shubik. Oxford: Oxford University Press, 2001.

———. *The Unity of Knowledge: The Convergence of the Natural and Human Science.* New York: New York Academy of Science, 2001.

Damasio, Antonio, et al. "Central Achromatopsia: Behavioral, Anatomic, and Physiological Aspects." *Neurology* 30: 1064–1071.

Edelman, Gerald. *Neural Darwinism: The Theory of Neuronal Group Selection.* New York: Basic Books, 1987.

———. "Neural Dynamics in a Model of the Thalamocortical System, 2: The Role of Neural Synchrony tested Through the Perturbations of Spike Timing." *Cerebral Cortex* 7 (1997): 228–236.

Edelman, Gerald, and Giulio Tononi. *A Universe of Consciousness: How Matter Becomes Imagination.* Cambridge: Perseus, 2000.

Engberg, P. "The Concept of Domain in Cognitive Theory of Metaphor." *Nordic Journal of Linguistics* 18: 111–119.

———. "Space and Time: A Metaphoric Relation." In *Cognitive Semantics: Meaning and Cognition.* Amsterdam: John Benjamin, 1999.

Feldman, Jerome. *From Molecule to Metaphor.* Cambridge, Mass.: MIT Press, 2006.

Gallese, V., and T. Metzinger. "The Emergence of a Shared Action Ontology: Building Blocks for a Theory." *Consciousness and Cognition* 12 (2003): 549–571.

Gardenfors, P. "Ambiguity, Harmony, and Probability." In *Odds and Ends: Philosophical Essays Dedicated to Wlodek Rabinwicz on the Occasion of His Fiftieth Birthday.* Uppsala: Uppsala Philosophical Studies, 1996.

———. *Conceptual Spaces: The Geometry of Thought.* Cambridge, Mass.: MIT Press, 2004.

———. "The Emergence of Meaning." *Linguistics and Philosophy* 16 (1993): 285–309.

———. "A Geometric Model of Concept Formation." In *Information Modeling and Knowledge Bases III.* Amsterdam: IOS, 1992.

———. "Induction and the Evolution of Perceptual Spaces." In *Charles S. Peirce and the Philosophy of Science.* Tuscaloosa: University of Alabama Press, 1993.

———. "The Pragmatic Role of Modality in Natural Language." In *The Role of Pragmatics in Contemporary Philosophy*. Vienna: Holder-Pichler-Tempsky, 1998.

Gibbs, R. "The Cognitive Psychological Reality of Image Schemas and Their Transformations." *Cognitive Linguistics* 6 (1995): 347–378.

Goodwin, Brian, and Richard Sole. *Signs of Life: How Complexity Pervades Biology*. New York: Basic Books, 2000.

Hahn, U. "Concepts and Similarity." In *Knowledge, Concepts, and Categories*. East Sussex: Psychology Press, 1997.

Holland, John. *Hidden Order: How Adaptation Builds Complexity*. Cambridge: Perseus, 1995.

Johnson, Mark. *The Body in the Mind*. Chicago: University of Chicago Press, 1987.

———. "Dewey's Zen: The 'Oh' of Wonder." Discussion paper at the Society for the Advancement of American Philosophy.

———. "The Philosophic Significance of Image Schemas." In *From Perception to Meaning: Image Schemas in Cognitive Linguistics*, ed. B. Hampe. New York: Mouton de Gruyter, 2005.

Johnson, Mark, and George Lakoff. *Philosophy in the Flesh: The Embodied Mind and Its Challenge to Western Thought*. New York: HarperCollins, 1999.

Kaag, John. "The Neural Dynamics of the Imagination." *Phenomenology and the Cognitive Sciences* 7, no. 4 (2008).

Kauffman, Stuart. *Investigations*. Oxford: Oxford University Press, 2000.

Kohler, E. "Hearing Sounds, Understanding Actions: Action Representation in Mirror Neurons." *Science* 297.

Lakoff, George. *Women, Fire, and Dangerous Things: What Categories Reveal about the Mind*. Chicago: University of Chicago Press, 1987.

LeDoux, Joseph. *The Emotional Brain: The Mysterious Underpinnings of Emotional Life*. New York: Touchstone, 1996.

———. *Synaptic Self: How Our Brains Become Who We Are*. New York: Viking, 2002.

Lewis-William, David. *The Mind in the Cave: Consciousness and the Origins of Art*. London: Thames and Hudson, 2002.

Llinas, Rudolfo. *I of the Vortex: From Neurons to Self*. Cambridge, Mass.: MIT Press, 2001.

Maturana, Humberto. "Autopoeisis: The Organization of the Living." In *Autopoeisis and Cognition*. Dordrecht: Reidel, 1980.

Maturana, Humberto, and Francisco Varela. *The Tree of Knowledge*. Boston: Shambhala, 1987.

Merleau-Ponty, Maurice. "Cezanne's Doubt." In *Sense and Non-sense*. Chicago: Northwestern University Press, 1964.

Mingers, John. *Self-Producing Systems*. New York: Plenum, 1995.

Ramachandran, V., and S. Blakeslee. *Phantoms in the Brain: Probing the Mysteries of the Human Brain*. New York: HarperCollins, 1998.

Rao, R., et al., eds. *Probabilistic Models of the Brain: Perception and Neural Function*. Cambridge, Mass.: MIT Press, 2002.

Rizzolatti, G., and L. Craighero. "The Mirror Neuron System." *Annual Review of Neuroscience* 27 (2004): 169–192.

Rohrer, Tim. "Image Schemata and the Brain." In *From Perception to Meaning: Image Schemas in Cognitive Linguistics*, ed. B. Hampe and J. Grady. Berlin: Mouton de Gruyter, 2006.

Rose, Susan. "Cross Modal Abilities in Human Infants." In *Handbook of Infant Development*, ed. J. Osofsky. New York: Wiley, 2001.

Shepard, R. "The Analysis of Proximities: Multidimensional Scaling with an Unknown Distance Function. II." *Psychometrika* 27 (1962): 219–246.

———. "Toward a Universal Law of Generalization for Psychological Science." *Science* 237 (1987): 1317–1323.

Sinha, Christopher. "Language, Cultural Context, and the Embodiment of Spatial Cognitions." *Cognitive Linguistics* 11, no. 2: 14–41.

Sporns, O., et al. "Modeling Perceptual Grouping and Figure-Ground Segregation by Means of Active Reentrant Connections." *Proceedings of the National Academy Science of the United States of America* 88 (1991): 129–133.

Tononi, G., et al. "Measures of Degeneracy and Redundancy in Biological Networks." *Proceedings of the National Academy of Sciences of the United States of America* 96 (1995): 3188–3208.

Tucker, Donald. *Mind from Body: Neural Structures of Experience*. Book manuscript. 2007.

Varela, Francisco. *The Embodied Mind: Cognitive Science and Human Experience*. Cambridge, Mass.: MIT Press, 1992.

Index

AMERICAN PHILOSOPHY
Douglas R. Anderson and Jude Jones, series editors

Kenneth Laine Ketner, ed., *Peirce and Contemporary Thought: Philosophical Inquiries.*

Max H. Fisch, ed., *Classic American Philosophers: Peirce, James, Royce, Santayana, Dewey, Whitehead, second edition.* Introduction by Nathan Houser.

John E. Smith, *Experience and God, second edition.*

Vincent G. Potter, *Peirce's Philosophical Perspectives.* Edited by Vincent Colapietro.

Richard E. Hart and Douglas R. Anderson, eds., *Philosophy in Experience: American Philosophy in Transition.*

Vincent G. Potter, *Charles S. Peirce: On Norms and Ideals, second edition.* Introduction by Stanley M. Harrison.

Vincent M. Colapietro, ed., *Reason, Experience, and God: John E. Smith in Dialogue.* Introduction by Merold Westphal.

Robert J. O'Connell, S.J., *William James on the Courage to Believe, second edition.*

Elizabeth M. Kraus, *The Metaphysics of Experience: A Companion to Whitehead's "Process and Reality," second edition.* Introduction by Robert C. Neville.

Kenneth Westphal, ed., *Pragmatism, Reason, and Norms: A Realistic Assessment—Essays in Critical Appreciation of Frederick L. Will.*

Beth J. Singer, *Pragmatism, Rights, and Democracy.*

Eugene Fontinell, *Self, God, and Immorality: A Jamesian Investigation.*

Roger Ward, *Conversion in American Philosophy: Exploring the Practice of Transformation.*

Michael Epperson, *Quantum Mechanics and the Philosophy of Alfred North Whitehead.*

Kory Sorrell, *Representative Practices: Peirce, Pragmatism, and Feminist Epistemology.*

Naoko Saito, *The Gleam of Light: Moral Perfectionism and Education in Dewey and Emerson.*

Josiah Royce, *The Basic Writings of Josiah Royce.*

Douglas R. Anderson, *Philosophy Americana: Making Philosophy at Home in American Culture.*

James Campbell and Richard E. Hart, eds., *Experience as Philosophy: On the World of John J. McDermott.*

John J. McDermott, *The Drama of Possibility: Experience as Philosophy of Culture.* Edited by Douglas R. Anderson.

Larry A. Hickman, *Pragmatism as Post-Postmodernism: Lessons from John Dewey.*

Larry A. Hickman, Stefan Neubert, and Kersten Reich, eds., *John Dewey Between Pragmatism and Constructivism.*

Dwayne A. Tunstall, *Yes, But Not Quite: Encountering Josiah Royce's Ethico-Religious Insight.*

Josiah Royce, *Race Questions, Provincialism, and Other American Problems, expanded edition.* Edited by Scott L. Pratt and Shannon Sullivan.

Lara Trout, *The Politics of Survival: Peirce, Affectivity, and Social Criticism.*

John R. Shook and James A. Good, *John Dewey's Philosophy of Spirit, with the 1897 Lecture on Hegel.*

Josiah Warren, *The Practical Anarchist: Writings of Josiah Warren.* Edited and with an Introduction by Crispin Sartwell.

Naoko Saito and Paul Standish, eds., *Stanley Cavell and the Education of Grownups.*

Douglas R. Anderson and Carl R. Hausman, *Conversations on Peirce: Reals and Ideals.*

Rick Anthony Furtak, Jonathan Ellsworth, and James D. Reid, eds., *Thoreau's Importance for Philosophy.*

James M. Albrecht, *Reconstructing Individualism: A Pragmatic Tradition from Emerson to Ellison.*

Mathew A. Foust, *Loyalty to Loyalty: Josiah Royce and the Genuine Moral Life.*

Cornelis de Waal and Krysztof Piotr Skowroński (eds.), *The Normative Thought of Charles S. Peirce.*

Dwayne A. Tunstall, *Doing Philosophy Personally: Thinking about Metaphysics, Theism, and Antiblack Racism.*

Erin McKenna, *Pets, People, and Pragmatism.*

Sami Pihlström, *Pragmatic Pluralism and the Problem of God.*

Thomas M. Alexander, *The Human Eros: Eco-ontology and the Aesthetics of Existence.*

John Kaag, *Thinking Through the Imagination: Aesthetics in Human Cognition.*

Kelly A. Parker and Jason Bell (eds.), *The Relevance of Royce.*

W. E. B. Du Bois, *The Problem of the Color Line at the Turn of the Twentieth Century: The Essential Early Essays.* Edited by Nahum Dimitri Chandler.

Nahum Dimitri Chandler, *X—The Problem of the Negro as a Problem for Thought.*